(LP) The LogicPrep Guide to Algebra

LogicPrep

About this book

This book is the culmination of many hours of diligent work by the following people: Jesse Kolber, Roger Reiersen, Jamie Kenyon, Helen Moss, Brian Siberine, Molly Pickel, Matthew Kwong, Chad Schwam, and Alex Wurm. Our dynamic team collectively has over 30 years in the test-prep industry and 20,000 hours of preparing students for both the ACT and SAT.

Published by LogicPrep Tutoring, 2018
ISBN: 978-0-9851060-3-4

Contents

A Note from LogicPrep

Hi there. Welcome to the high-stakes, much-dreaded, your-whole-life-seems-to-hang-in-the-balance world of college entry exams.

Here's the good news:

If you're reading this book, you're already smart – not because you chose our book (well, not *just* because), but because you've realized that the best way to get the ACT or SAT score that you want is to prepare. Prepare logically, even.

To be sure, your ACT or SAT score will be only a part of your entire college application. Your grades, extracurriculars, outside activities, volunteer work, and of course your main essay and supplements will all factor in. But let's be honest: your test scores *will* be a big part of your application.

How did this come to be, you ask? Why are there standardized tests like the ACT and SAT in the first place? There are many reasons, but the biggest one is basically that the American education system is set up on a district-by-district basis – so colleges need a uniform national standard to find out how much you know and where you fit in with your peers. Thus, the ACT and the SAT test how much you know (of what the test-makers deem important, that is) and where your score fits in on a national bell curve. Simple, right?

This book is organized to teach you clearly and thoroughly the primary algebra skills that you'll need on the math sections of the ACT and SAT. After you master these skills, you'll need to learn how the ACT and SAT expect you to apply them. And, even more importantly, you'll need – through thorough reflection and rigorous reflection – to learn how you best take the test.

We're not in the testing room with you, but we will offer you our wisdom, honed from years of working with all different kinds of students, so we can alert you to common (and sometimes not-so-common) mistakes… and some variations on test-taking strategy so that you can find what works best for you.

Finally, after you learn all the concepts the ACT and SAT test and identify the test strategies that work best for you, there are three more skills (which may not sound like skills) that you'll need to master: being confident, being calm, and being careful.

- **BE CONFIDENT** – You have learned the knowledge you need and have seen how the ACT and SAT test it, so be strong! Neither the ACT nor the SAT varies much from test to test.
- **BE CALM** – Becoming emotional on these tests will work against you. Whether you feel you are performing well or not, you must stay calm instead of becoming excited or nervous.
- **BE CAREFUL** – Even experienced test-takers can be sloppy, or can misread a question, or can forget to double check exactly what the question is asking for… and so they let themselves get caught off-guard by a well-written but incorrect answer choice.

The road to success will be a little different for everyone, and the exact final destinations will vary, too. But here's the promise we make:

If you work hard to learn the skills we teach, and if you practice and adjust your test-taking skills, your score will improve – often by a lot. How much improvement you attain depends on *you*: on *your* hard work and on *your* commitment.

Lastly, we do teach these skill sets so you can do great on the ACT and SAT, but we wouldn't be in this business if we didn't also believe we were teaching you things that would be vital for the rest of your life.

So jump in. The water's fine. Have some fun and know you are making an investment that will pay off – on the ACT and SAT, and beyond!

– The LogicPrep Team

Algebraic Operations
Algebra and Arithmetic Problem Set 1

1. If $6x - 3 = 10x + 5$, then $x =$?

    A. -2

    B. $-\dfrac{1}{2}$

    C. 0

    D. $\dfrac{1}{2}$

    E. 2

2. If $15x = -5(8 - x)$, then $x =$?

    F. -4

    G. -2

    H. $\dfrac{10}{4}$

    J. $\dfrac{20}{7}$

    K. 4

3. $(2x^2 - 6x + 7) - (x^2 - 12x - 3)$ is equivalent to:

    A. $x^2 - 18x + 4$

    B. $x^2 + 6x + 4$

    C. $x^2 + 6x + 10$

    D. $3x^2 + 18x + 10$

    E. $3x^2 - 18x + 4$

4. If $|x - 2| \geq 10$, which is a possible value for x?

    F. -10

    G. -7

    H. -2

    J. 2

    K. 10

DO YOUR FIGURING HERE

Algebraic Operations
Algebra and Arithmetic Problem Set 1

5. If $3c + 6 < 9$, which of the following cannot be c?

    A. -1

    B. 0

    C. $\dfrac{1}{2}$

    D. $\dfrac{9}{10}$

    E. $\dfrac{10}{9}$

DO YOUR FIGURING HERE

6. If $2^{3x} = 64$, $x =$

    F. -2

    G. 0

    H. $\dfrac{1}{2}$

    J. 1

    K. 2

7. If $\dfrac{x}{9} = \dfrac{2x}{c}$ and $x \neq 0$, what is the value of c?

    A. 1

    B. 2

    C. 4.5

    D. 9

    E. 18

8. If $(x - 4)^3 = -125$, what is the value of x?

    F. -9

    G. -5

    H. -1

    J. 5

    K. 9

9. If k is a constant and $4x + 5 = 2kx + 5$ for all values of x, what is the value of k?

  A. -2

  B. 0

  C. 1

  D. 2

  E. 5

10.
$$\frac{(4 - x)}{3} = \frac{7}{2}$$

What number, when used as x above, makes the statement true?

  F. -7.5

  G. -6.5

  H. 3

  J. 6.5

  K. 7.5

11. If $3z + z = 5z - 1$, $z =$

  A. -1

  B. 1

  C. 2

  D. 3

  E. 4

12.
$$4x^3 > (4x)^3$$

Which value of x makes the above statement false?

  F. -4

  G. -1

  H. $\dfrac{-1}{2}$

  J. $\dfrac{1}{2}$

  K. No value of x

13. If $n = 4z$, for what value of z is $n = z$?

    A. 0

    B. $\dfrac{1}{4}$

    C. 1

    D. 4

    E. No value of z

DO YOUR FIGURING HERE

14. The expression $[a - (b + c)]\, d$ is equivalent to:

    F. $ad - bd + cd$

    G. $ad - bd - cd$

    H. $a - bd + cd$

    J. $a - bd - cd$

    K. $ad - b + c$

15. For all positive integers, a, b, and c, which of the following expressions is equivalent to $\dfrac{b}{a}$?

    A. $\dfrac{b + c}{a + c}$

    B. $\dfrac{b - c}{a - c}$

    C. $\dfrac{b \cdot b}{a \cdot a}$

    D. $\dfrac{b \cdot a}{a \cdot b}$

    E. $\dfrac{b \cdot c}{c \cdot a}$

16. If $x = -(y + 3)$, then $(x + y)^3 = ?$

    F. -27

    G. -9

    H. -3

    J. 9

    K. 27

Algebraic Operations
Algebra and Arithmetic Problem Set 1

7. $\left(a - \dfrac{1}{3}b\right)^2 = ?$

  A. $a^2 - \dfrac{1}{9}b^2$

  B. $a^2 - \dfrac{2}{3}ab + \dfrac{1}{9}b^2$

  C. $a^2 - \dfrac{2}{3}ab + \dfrac{1}{3}b^2$

  D. $a^2 + b^2$

  E. $a^2 - \dfrac{1}{3}ab - \dfrac{1}{9}b^2$

8. When $x - y = 8$, what is the value of

$3(x - y) + (x - y)^2 - \dfrac{x - y}{4} - 15?$

  F. 15

  G. 55

  H. 71

  J. 72

  K. 101

9. If $x = 3a + 5$ and $y = 6 - a$, which of the following expresses x in terms of y?

  A. $x = 3y - 13$

  B. $x = \dfrac{y - 11}{3}$

  C. $x = \dfrac{y - 23}{3}$

  D. $x = 3y + 15$

  E. $x = 23 - 3y$

Answer Key

#	Answer	Frequency	Difficulty
1	A	popular	1
2	F	popular	1
3	C	popular	1
4	F	popular	1
5	E	popular	1
6	K	popular	1
7	E	popular	1
8	H	popular	1
9	D	popular	1
10	G	popular	1
11	B	popular	1
12	J	popular	2
13	A	popular	2
14	G	popular	2
15	E	popular	2
16	F	popular	3
17	B	popular	2
18	H	popular	2
19	E	popular	3

Fractions
Quick Drill

1. Simplify the fraction:
$\dfrac{21}{49}$

2. Simplify the fraction:
$\dfrac{99}{12}$

3. $\dfrac{1}{2} + \dfrac{1}{6} =$

4. $\left(\dfrac{1}{2}\right)\left(\dfrac{1}{6}\right) =$

5. $\dfrac{\frac{1}{2}}{\frac{1}{6}} =$

6. $\dfrac{2}{5}\left(\dfrac{4}{3}\right) =$

7. $\dfrac{3}{8} + \dfrac{1}{2} + \dfrac{2}{3} - \dfrac{4}{9} =$

8. $3\dfrac{1}{2} + 4\dfrac{1}{4} =$

9. $1\dfrac{1}{2}\left(3\dfrac{1}{3}\right) =$

10. $\dfrac{\frac{1}{2} - 13}{\frac{2}{8}} =$

Fractions
Quick Drill

11. $\dfrac{72}{14} + \dfrac{3}{21} - \dfrac{3}{42} =$

12. $\dfrac{15}{13}\left(\dfrac{2}{3}\right) =$

13. What is $\dfrac{1}{2}$ of $\dfrac{8}{9}$?

14. What is $\dfrac{1}{3}$ of 120?

15. $\dfrac{\frac{2}{3}}{\frac{7}{8}} =$

16. $\dfrac{\frac{3}{4}}{\frac{5}{8}} - \dfrac{\frac{7}{15}}{\frac{1}{3}} =$

17. $\dfrac{7}{2} + \dfrac{2}{7} + \dfrac{5}{6} =$

18. $\left(\dfrac{17}{3}\right)\left(\dfrac{3}{18}\right) =$

19. $\left(\dfrac{36}{8}\right)\left(\dfrac{16}{6}\right) =$

20. $\dfrac{5}{8} + \dfrac{1}{3} - \dfrac{1}{12} + \dfrac{5}{6} =$

Fractions
Quick Drill

21. $62\left(\dfrac{5}{9}-\dfrac{1}{2}\right)=$

22. What is one eighth of $6\dfrac{3}{4}$?

23. What is the sum of $\dfrac{14}{15}$ and $\dfrac{1}{2}$?

24. $\left(\dfrac{3}{4}-\dfrac{1}{8}\right)\left(\dfrac{1}{6}+\dfrac{2}{5}\right)=$

25. $4\left(\dfrac{\frac{3}{8}}{\frac{6}{7}}\right)=$

26. $\dfrac{8}{3}\left(7+\dfrac{1}{4}\right)=$

27. $2x^2\left(\dfrac{\frac{1}{2}+\frac{1}{3x}}{\frac{2x}{5}}\right)=$

28. $\dfrac{2x}{3}\left(\dfrac{5}{8x}+\dfrac{1}{4}\right)=$

29. $\dfrac{1}{4x}-\dfrac{1}{8x}=$

30. $\left(\dfrac{4x+5}{2}\right)-\left(\dfrac{2x+1}{3}\right)=$

Fractions
Quick Drill

Answer Key

#	Answer
1	$\dfrac{3}{7}$
2	$\dfrac{33}{4}$ or $8\dfrac{1}{4}$
3	$\dfrac{2}{3}$
4	$\dfrac{1}{12}$
5	3
6	$\dfrac{8}{15}$
7	$\dfrac{79}{72}$ or $1\dfrac{7}{72}$
8	$\dfrac{31}{4}$ or $7\dfrac{3}{4}$
9	5
10	-50
11	$\dfrac{73}{14}$ or $5\dfrac{3}{14}$
12	$\dfrac{10}{13}$
13	$\dfrac{4}{9}$
14	40
15	$\dfrac{16}{21}$
16	$-\dfrac{1}{5}$
17	$\dfrac{97}{21}$ or $4\dfrac{13}{21}$
18	$\dfrac{17}{18}$
19	12
20	$\dfrac{41}{24}$ or $1\dfrac{17}{24}$
21	$\dfrac{31}{9}$ or $3\dfrac{4}{9}$
22	$\dfrac{27}{32}$
23	$\dfrac{43}{30}$ or $1\dfrac{13}{30}$
24	$\dfrac{17}{48}$
25	$\dfrac{7}{4}$

#	Answer
26	$\dfrac{58}{3}$ or $19\dfrac{1}{3}$
27	$\dfrac{15x+10}{6}$
28	$\dfrac{5+2x}{12}$
29	$\dfrac{1}{8x}$
30	$\dfrac{8x+13}{6}$

Solving for One Variable
Quick Drill

1. Solve for x:
$7x = 21$

2. Solve for m:
$m + 2 = 18$

3. Solve for c:
$3c + 12 = \dfrac{4}{5}$

4. Solve for x:
One eighth of x is 16.

5. Solve for x:
$\dfrac{2}{x} = 14$

6. Solve for r:
$r^2 = 49$

7. Solve for y:
$2^y = 8$

8. Solve for z:
The square root of z is equal to 3.

9. Solve for k:
$4(k - 2) = 28$

10. Solve for x:
$x\left(4 + \dfrac{1}{2}\right) = 9$

Solving for One Variable
Quick Drill

11. Solve for n:
$$\frac{n+2}{4} = \frac{3n}{8}$$

12. Solve for x:
$$x(x+8) = 0$$

13. Solve for b:
One half of b is the same as the sum of b and 3.

14. Solve for x:
$$\frac{13x}{2} = 26$$

15. Solve for k:
$$9(k+2k) = 12$$

16. Solve for c:
$$(3+c)\left(\frac{1}{2} - \frac{3}{8}\right) = 2$$

17. Solve for b:
$$\frac{\left(\frac{b}{2}\right)}{\left(\frac{4}{3}\right)} = 1$$

18. Solve for x:
$$\frac{x}{\left(\frac{2}{3}\right)} = 4(x+3)$$

19. Solve for x:
$$3x - 7 = \frac{2(x+2)}{9}$$

20. Solve for m:
$$3\left(m - \frac{4}{5}\right) = \frac{m}{3}$$

Solving for One Variable
Quick Drill

Answer Key

#	Answer
1	$x = 3$
2	$m = 16$
3	$c = -\dfrac{56}{15}$ or $-3\dfrac{11}{15}$
4	$x = 128$
5	$x = \dfrac{1}{7}$
6	$x = \pm 7$
7	$y = 3$
8	$z = 9$
9	$k = 9$
10	$x = 2$
11	$n = 4$
12	$x = 0, -8$
13	$b = -6$
14	$x = 4$
15	$k = \dfrac{4}{9}$
16	$c = 13$
17	$b = \dfrac{8}{3}$
18	$x = -\dfrac{24}{5}$ or $-4\dfrac{4}{5}$
19	$x = \dfrac{67}{25}$ or $2\dfrac{17}{25}$
20	$m = \dfrac{9}{10}$

Solving for Two Variables
Quick Drill

If $4x + 2y = 8$ and $y - 3 = 9$ then what are the values of x and y?

6. If $7x + 3y = 24$ and $\dfrac{x + y}{4} = 22$ then what are the values of x and y?

If $\dfrac{a}{2} = 6$ and $2ab = 12$ then what is the value of $\dfrac{a}{b}$?

7. If $17 + a + 2b = 35$ and $\dfrac{a}{b} = 4$ then what is the value of a - b?

If $2m - 5 = n$ and $m + 2n = 10$ then what are the values of m and n?

8. If $\dfrac{a}{b + 4} = \dfrac{6}{8}$ and $\dfrac{a}{2} = \dfrac{b}{3}$ then what are the values of a and b?

If $7y + 2 = 16$ and $2x + 3y = 18$ then what is the value of $x + y$?

9. If $\dfrac{x}{2y} = 3$ and $\dfrac{x}{2} - y + 2 = 18$ then what is the value of $\dfrac{2x}{y}$?

If $4j = 3k - 1$ and $j + \dfrac{k}{2} = 16$ then what are the values of j and k?

10. Solve for m and n:
If $m + \dfrac{3n}{8} = 2m - 5$ and $\dfrac{m}{6} = n$ then what are the values of m and n?

Solving for Two Variables
Quick Drill

11. Solve for j and k:

$3j - 2k = k + 6$ and $\dfrac{j}{k} = \dfrac{2}{3}$

12. Solve for g and h:

$14g + 16h + 2 = 10g + 4h$ and $3h = 2h + \text{(}$

13. Express k in terms of y:

$y = kx$ and $x + 3y = 7$

14. Solve for x and y:

$\dfrac{x}{5} = y - 10$ and $4x - 3 = 2y + 1$

15. Solve for ab:

$a = b - 3a$ and $-\dfrac{a}{8} + \dfrac{b}{2} = \dfrac{3}{4}$

16. If $x^2 - 9y^2 = 15$ and $x + 3y = 5$, then what are the values of x and y?

17. Solve for all values of g and h:

$g^2 + 2 = 6$ and $g + h = 3$

18. Solve for a and b:

$\dfrac{b}{a} = \dfrac{16}{3} + 2$ and $a + \dfrac{b}{2} = 12$

19. Solve for x and y:

$\dfrac{2x + y}{3} = \dfrac{x - y}{2}$ and $x - 4y = 3$

20. If $2x + 3y = 0$ and $x + y = 6$ then what are the values of x and y?

Solving for Two Variables
Quick Drill

Answer Key

#	Answer
1	$x = -4$ and $y = 12$
2	$\dfrac{a}{b} = 24$
3	$m = 4$ and $n = 3$
4	$x + y = 8$
5	$j = \dfrac{19}{2}$ and $k = 13$
6	$x = -60$ and $y = 148$
7	$a - b = 9$
8	$a = -24$ and $b = -36$
9	$\dfrac{2x}{y} = 12$
10	$m = \dfrac{16}{3}$ and $n = \dfrac{8}{9}$
11	$j = -4$ and $k = -6$
12	$g = -\dfrac{1}{8}$ and $h = -\dfrac{1}{8}$
13	$k = \dfrac{y}{7 - 3y}$
14	$x = \dfrac{20}{3}$ and $y = \dfrac{34}{3}$
15	$ab = \dfrac{16}{25}$
16	$x = 4$ and $y = \dfrac{1}{3}$
17	$g = \pm 2$ and $h = 1$ or 5
18	$a = \dfrac{18}{7}$ and $b = \dfrac{132}{7}$
19	$x = \dfrac{5}{3}$ and $y = -\dfrac{1}{3}$
20	$x = 18$ and $y = -12$

Systems of Equations
Quick Drill

1. Solve for x and y:
$3x + 2y = 6$
$-2x - 2y = 3$

2. Solve for x and y:
$14x + 3y = 6$
$14x - 2y = 3$

3. Solve for x and y:
$2x - y = 5$
$-x + 3y = 5$

4. Solve for $3x - 2y$:
$2x + 9y = 16$
$x - 11y = 14$

5. Solve for x and y:
$x + 3y = 16$
$y - 8x = -78$

6. Solve for x and y:
$5x + 3y = 7$
$3x - 5y = -23$

7. Solve for x and y:
$6x + 2y = 7$
$y - 4x = 0$

8. Solve for x and y:
$4x + 8y = 20$
$-4x + 2y = 0$

9. Solve for $2x + 4y$:
$6x - 3y = 9$
$4x - 7y = -3$

10. Solve for x and y:
$2x + 3y = 11$
$4x - 6y = 18$

Systems of Equations
Quick Drill

11. Solve for x and y:
$$7x + 2y = -18$$
$$3x - y = -4$$

12. Solve for x and y:
$$10x - 3y = 58$$
$$4x + 2y = 20$$

13. Solve for x and y:
$$x^2 - 4y = 7$$
$$x + 2y = 4$$

14. Solve for x and y:
$$x^2 - y^2 = 21$$
$$x + y = 3$$

15. Solve for x and y:
$$5x + 7y = -25$$
$$11x + 6y = -8$$

16. The system of equations below has a solution at $x = 5$. What is the value of k?
$$3x + \frac{1}{2}y = 15$$
$$kx - 2y = 10$$

17. What is the value of x in the following systems of equations, in terms of c and d?
$$3x + y = c$$
$$x - y = d$$

18. For what value of c would the system of equations have infinitely many solutions?
$$3x + 4y = 8$$
$$9x + 12y = 2c$$

19. For what value of a would the system of equations have infinitely many solutions?
$$x - 4y = 8$$
$$2y - \frac{x}{2} = 2a$$

20. For what value of k would the system of equations have no solutions?
$$2x - 3y = 5$$
$$3x + ky = 4$$

Systems of Equations
Quick Drill

Answer Key

#	Answer
1	$x = 9$ and $y = -\dfrac{21}{2}$
2	$x = \dfrac{3}{10}$ and $y = \dfrac{3}{5}$
3	$x = 4$ and $y = 3$
4	$3x - 2y = 30$
5	$x = 10$ and $y = 2$
6	$x = -1$ and $y = 4$
7	$x = .5$ and $y = 2$
8	$x = 1$ and $y = 2$
9	$2x + 4y = 12$
10	$x = 5$ and $y = \dfrac{1}{3}$
11	$x = -2$ and $y = -2$
12	$x = 5.5$ and $y = -1$
13	$x = -5$ and $y = \dfrac{9}{2}$ OR $x = 3$ and $y = \dfrac{1}{2}$
14	$x = 5$ and $y = -2$
15	$x = 2$ and $y = -5$
16	$k = 2$
17	$x = \dfrac{c + d}{4}$
18	$c = 12$
19	$a = -2$
20	$k = -\dfrac{9}{2}$

Systems of Equations
Algebra and Arithmetic Problem Set 2

1. There were 800 people who attended a school musical. Students sold $5,800 worth of tickets in all. An adult ticket costs $8 and a student ticket costs $6. How many adults went?

   A. 0

   B. 300

   C. 400

   D. 500

   E. 800

DO YOUR FIGURING HERE

2. On Monday, Jeff buys 2 hotdogs and a hamburger for 13 dollars. On Tuesday, Jeff buys 3 hotdogs and 2 hamburgers for 21 dollars. How much money would a hotdog and a hamburger cost?

   F. 2

   G. 3

   H. 5

   J. 8

   K. 11

3. A theater sells popcorn and nachos. The theater made $172 when 13 people bought popcorn and 24 people bought nachos. They made $161 when 24 people bought popcorn and 13 people bought nachos. How much are the nachos?

   A. $4

   B. $4.50

   C. $5

   D. $6

   E. $10

4. A cafe has 13 tables that can seat a total of 44 people. Some of the tables seat 2 people and the rest seat 4 people. How many tables seat 4 people?

   F. 4

   G. 6

   H. 7

   J. 9

   K. 10

Systems of Equations
Algebra and Arithmetic Problem Set 2

5. Roger collected 100 dimes and nickels worth a total of $6.00. How many dimes did Roger collect?

    A. 20

    B. 25

    C. 30

    D. 50

    E. 80

6. Tickets for a movie cost $6.00 for a child and $10.00 for an adult. If 300 tickets were sold for a total of $2,800, what was the ratio of the number of children's tickets sold to the number of adults' tickets sold?

    F. 1 to 4

    G. 1 to 5

    H. 3 to 1

    J. 4 to 1

    K. 5 to 1

7. The cost of a cab ride using company F is $4.00 for any distance up to and including 2 miles and $2.50 per mile thereafter. The cost of a cab ride using company G is $2.45 per mile for any distance. For a ride that lasts m miles, the cost using company F is the same as the cost using company G. If m is a positive integer greater than 2, what is the value of m?

8. Two green balls and one red ball weigh an average of $6\frac{1}{3}$ pounds. Two red balls and one green ball weigh an average of $7\frac{2}{3}$ pounds. How much does a green ball weigh?

    F. 5 lbs

    G. 6.7 lbs

    H. 7 lbs

    J. 9 lbs

    K. 11 lbs

DO YOUR FIGURING HERE

Systems of Equations
Algebra and Arithmetic Problem Set 2

9. GetSwoll, a gym, charges each gym member $100 to join, plus $30 per month for membership. A rival gym, TuffEnuff, charges each member $50 to join, plus $35 per month for membership. A person who joins GetSwoll will pay the same total amount for membership as a person who joins TuffEnuff if each remains a member for how many months?

A. 4

B. 6

C. 8

D. 10

E. 12

DO YOUR FIGURING HERE

10. A bus of 30 campers from Camp Getaway stopped at the Snack Shack for hamburgers and hot dogs. The Snack Shack sells hamburgers for $4 and hot dogs for $3. If each camper got either one hamburger or one hot dog and the meals cost a total of $103, how many campers got a hamburger?

F. 10

G. 13

H. 17

J. 20

K. 21

11. John baked pies for his friends and family this past Thanksgiving. It takes him 40 minutes to bake an apple pie and 20 minutes to bake a pumpkin pie. If he spends a total of 8 hours baking pies, one at a time, and bakes twice as many pumpkin pies as apple pies then how many pumpkin pies did he bake?

A. 2

B. 3

C. 4

D. 6

E. 12

Systems of Equations
Algebra and Arithmetic Problem Set 2

12. Willy's window company received an order for two types of windows: Large panoramic and Standard bedroom. This order consists of a total of 24 windows and there are twice as many Standard bedroom windows as there are Large panoramic. How many Standard windows were in this order?

   F. 4

   G. 6

   H. 8

   J. 16

   K. 20

13. On a recent exam, some questions were worth 3 points each and the rest were worth 5 points each. Kristaps correctly answered the same number of 3-point questions as 5-point questions and scored an 88 on the exam. How many 5-point questions did he answer correctly?

   A. 20

   B. 16

   C. 13

   D. 11

   E. 10

14. The senior class at Ridgewood High School sponsored a dinner dance. Each student who helped prepare the food paid $3 admission, and each student who did NOT help prepare food paid $12 admission. A total of $1,188 was collected from the 120 students who paid admission. How much of the amount was collected from students who did NOT help prepare food?

   F. $84

   G. $276

   H. $594

   J. $1,080

   K. $1,104

DO YOUR FIGURING HERE

Systems of Equations
Algebra and Arithmetic Problem Set 2

Answer Key

#	Answer	Frequency	Difficulty
1	D	popular	3
2	J	popular	3
3	C	popular	4
4	J	popular	4
5	A	popular	3
6	G	popular	3
7	20	average	3
8	F	popular	5
9	D	popular	3
10	G	popular	3
11	E	popular	3
12	J	popular	3
13	D	popular	3
14	K	popular	3

Intro to Distribution and Factoring
Quick Drill

1. Distribute:
$a\left(b + cd\right)$

2. Distribute:
$6\left(4y + 3\right)$

3. Distribute:
$2x\left(\dfrac{4y + 3}{5x}\right)$

4. Distribute:
$\dfrac{a}{3}\left(b + \dfrac{1}{4}\right)$

5. Distribute:
$\dfrac{6\left(x + y\right)}{2\left(3x + 1\right)}$

6. Factor:
$16x + 8$

7. Factor:
$42y - 14$

8. Factor:
$9d^2 + 3d$

9. Simplify:
$\dfrac{54x + 21}{3}$

10. Simplify:
$\dfrac{24b + 40}{4} + \dfrac{36b - 24}{3}$

11. If $\dfrac{b^8}{b^5} = 8$, then what is the value of b?

16. If $3x + y = 9$ and $27x^2 - 3y^2 = 45$ then what is the value of $3x - y$?

12. If $\dfrac{x^2 - 2x - 3}{x + 1} = 17$, then what is the value of $x - 3$?

17. If $\dfrac{x^2 + xy}{4x + 4y} = 12$ then what is the value of x?

13. If $a + 4b = 8$ and $a^2 + 6ab + 8b^2 = 24$, then what is the value of $a + 2b$?

18. If $2\left(\dfrac{c}{2} + d\right) = 3$ then what is the value of $c^2 + 4cd + 4d^2$?

14. If $\dfrac{j^2 k + 3j^3 k}{j^2 k} + \dfrac{j^3 k}{2jk} = 1$ then what is the value of j?

19. If $\dfrac{3x + 6xy}{12xy} = 2$ then what is the value of y?

15. If $r + 2s = 11$ then what is the value of $r^2 + 4rs + 4s^2$?

20. If $2f^2 + 8fg + 8g^2 = 98$ then what is the value of $f + 2g$?

Intro to Distribution and Factoring
Quick Drill

Answer Key

#	Answer
1	$ab + acd$
2	$24y + 18$
3	$\dfrac{8y + 6}{5}$
4	$\dfrac{ab}{3} + \dfrac{a}{12}$
5	$\dfrac{3x + 3y}{3x + 1}$
6	$8\left(2x + 1\right)$
7	$14\left(3y - 1\right)$
8	$3d\left(3d + 1\right)$
9	$18x + 7$
10	$2\left(9b + 1\right)$
11	$b = 2$
12	$x - 3 = 17$
13	$a + 2b = 3$
14	$j = 0, -6$
15	$r^2 + 4rs + 4s^2 = 121$
16	$3x - y = \dfrac{15}{9}$
17	$x = 48$
18	$c^2 + 4cd + 4d^2 = 9$
19	$y = \dfrac{1}{6}$
20	$f + 2g = \pm 7$

Distributing and Factoring Polynomials
Quick Drill

1. Distribute:
   $(a + 4)(a - 2) =$

2. Distribute:
   $r(r - 8)(r + 3) =$

3. Distribute:
   $s(t - u)(t + u) =$

4. Distribute:
   $(b + 4)^2 =$

5. Distribute:
   $(x - 3)^3 =$

6. Solve for x:
   $x^2 + 8x + 16 = 0$

7. Solve for a:
   $a^2 + 7a - 30 = 0$

8. Solve for b:
   $b^2 - 6b - 7 = 0$

9. Solve for f:
   $f^2 + 8f - 33 = 0$

10. Solve for f:
    $j^2 + 10j - 20 = 4$

Distributing and Factoring Polynomials
Quick Drill

11. If $x^2 + y^2 = 19$ and $xy = 3$ then what is the value of $(x + y)^2$?

12. If $\dfrac{p^2 - q^2}{(p - q)^2} = 13$ then what is the value of $\dfrac{p - q}{p + q}$?

13. Solve for g:
$3g^2 + 18g + 27 = 0$

14. Solve for x:
$2x^2 - 2x = 24$

15. Solve for m:
$6m^2 + 6m + 15 = 27$

16. Solve for x:
$2x^4 = 2x$

17. Solve for a:
$3a^3 + 6a^2 = 24a$

18. Solve for b:
$b^2 - 8b + 18 = 2$

19. Solve for z:
$9z(z + 4) + 50 = 14$

20. Solve for j:
$4j(j - 5) = 96$

Distributing and Factoring Polynomials
Quick Drill

Answer Key

#	Answer
1	$a^2 + 2a - 8$
2	$r^3 - 5r^2 - 24r$
3	$st^2 - su^2$
4	$b^2 + 8b + 16$
5	$x^3 - 9x^2 + 27x - 27$
6	$x = -4$
7	$a = -10, 3$
8	$a = -1, 7$
9	$f = -11, 3$
10	$j = -12, 2$
11	$(x + y)^2 = 25$
12	$\dfrac{p - q}{p + q} = \dfrac{1}{13}$
13	$g = -3$
14	$x = -3, 4$
15	$m = -2, 1$
16	$m = 0, 1$
17	$a = -4, 0, 2$
18	$b = 4$
19	$z = -2$
20	$j = -3, 8$

Common Algebra Mistakes
Algebra and Arithmetic Problem Set 3

DO YOUR FIGURING HERE

1. $3a^5 \cdot 5a^3$ is equivalent to:

   A. $8a^2$

   B. $8a^8$

   C. $8a^{15}$

   D. $15a^8$

   E. $15a^{15}$

2. The expression $\dfrac{6 + \frac{1}{4}}{2 + \frac{1}{8}}$ is equal to:

   F. $1\dfrac{7}{17}$

   G. $2\dfrac{15}{17}$

   H. $2\dfrac{16}{17}$

   J. 3

   K. $3\dfrac{7}{2}$

3. Consider all products xy such that x is divisible by 4 and y is divisible by 12. Which of the following whole number is NOT a factor of each product xy?

   A. 3

   B. 8

   C. 10

   D. 12

   E. 48

4. $|8 - 5| - |5 - 11| = ?$

   F. -9

   G. -3

   H. 3

   J. 9

   K. 29

5. The function f is defined by $f(x) = 3x^2 - 4x$.
What is the value of $f(-2)$?

   A. -20

   B. -8

   C. -4

   D. 4

   E. 20

DO YOUR FIGURING HERE

6. The expression $b^2 - 4b + 4$ is equivalent to:

   F. $(b+2)(b-2)$

   G. $(b-4)(b+\frac{1}{4})$

   H. $(b+4)(b-\frac{1}{4})$

   J. $(b+2)^2$

   K. $(b-2)^2$

7. The inequality $2x - 7 < x + 14$ is equivalent to
which of the following inequalities?

   A. $x < 5$

   B. $x < 9$

   C. $x < 10$

   D. $x < 21$

   E. $x < 42$

8. What is the sum of the 4 binomials listed below?
$2x^2 - 3x, 5x + 4, 3x - 8, x^2 + 3$

   F. $x^2 + 5x + 1$

   G. $x^2 + 5x + 5$

   H. $2x^2 + 8x + 1$

   J. $3x^2 + 5x - 1$

   K. $3x^2 + 5x + 7$

Common Algebra Mistakes
Algebra and Arithmetic Problem Set 3

9. Let $a < b < c < d < 0$ be true for integers a, b, c, and d. Which of the following expressions has the greatest value?

    A. $\dfrac{d}{a}$

    B. $\dfrac{b}{c}$

    C. $\dfrac{d}{c}$

    D. $\dfrac{a}{b}$

    E. $\dfrac{a}{d}$

10. For functions f and g defined by $f(x) = 3x^2 - x$ and $g(x) = 4x + 2$, what is the value of $f(g(2))$?

    F. 20

    G. 42

    H. 90

    J. 102

    K. 290

11. Which of the following inequalities is an equivalent algebraic expression for the statement below?

3 less than the product of 5 and a number a is less than 17

    A. $3 - 5a < 17$

    B. $5a - 3 < 17$

    C. $20 < 5a$

    D. $14 < 5a$

    E. $5a < 14$

12. How many minutes would it take a car to travel 60 miles at a constant speed of 100 miles per hour?

    F. 24

    G. 36

    H. 60

    J. 96

    K. 100

Common Algebra Mistakes
Algebra and Arithmetic Problem Set 3

13. When $x = \dfrac{1}{3}$, what is the value of $\dfrac{6x - 3}{x}$?

   A. -3

   B. -1

   C. $-\dfrac{1}{3}$

   D. $\dfrac{1}{3}$

   E. 3

14. If $x^a \cdot x^b = x^7$ for all x, which of the following must be true?

   F. $a - b = 7$

   G. $a + b = 7$

   H. $a \div b = 7$

   J. $a \cdot b = 7$

   K. $\sqrt{ab} = 7$

15. The value of x that will make $\dfrac{x}{4} + 2 = \dfrac{3}{4}$ a true statement lies between which of the following numbers?

   A. -6 and -2

   B. -2 and 0

   C. 0 and 2

   D. 2 and 3

   E. 3 and 6

16. Which of the following is the least common denominator for the expression below?

$$\frac{1}{7^6 \cdot 15 \cdot 19^2} + \frac{1}{7^3 \cdot 19^4} + \frac{1}{7^4 \cdot 15^5}$$

   F. $7 \cdot 15 \cdot 19$

   G. $7^6 \cdot 15 \cdot 19$

   H. $7^6 \cdot 15^5 \cdot 19^4$

   J. $7^6 \cdot 15^6 \cdot 19^4$

   K. $7^6 \cdot 15^6 \cdot 19^6$

DO YOUR FIGURING HERE

Common Algebra Mistakes
Algebra and Arithmetic Problem Set 3

17. The expression $[a - (b + c)]\, d$ is equivalent to:

DO YOUR FIGURING HERE

    A. $ad - bd + cd$

    B. $ad - bd - cd$

    C. $a - bd + cd$

    D. $a - bd - cd$

    E. $ad - b + c$

18. If $6x - 3 = 10x + 5$, then $x =?$

    F. -2

    G. $-\dfrac{1}{2}$

    H. 0

    J. $\dfrac{1}{2}$

    K. 2

19. For all positive integers, $a, b,$ and c, which of the following expressions is equivalent to $\dfrac{b}{a}$?

    A. $\dfrac{b + c}{a + c}$

    B. $\dfrac{b - c}{a - c}$

    C. $\dfrac{b \cdot b}{a \cdot a}$

    D. $\dfrac{b \cdot a}{a \cdot b}$

    E. $\dfrac{b \cdot c}{c \cdot a}$

20. If $x = -(y + 3)$, then $(x + y)^3 =?$

    F. -27

    G. -9

    H. -3

    J. 9

    K. 27

Common Algebra Mistakes
Algebra and Arithmetic Problem Set 3

21. If $15x = -5\,(8 - x)$, then $x = ?$

    A. -4

    B. -2

    C. $\dfrac{10}{4}$

    D. $\dfrac{20}{7}$

    E. 4

DO YOUR FIGURING HERE

22. $\left(2x^2 - 6x + 7\right) - \left(x^2 - 12x - 3\right)$ is equivalent to:

    F. $x^2 - 18x + 4$

    G. $x^2 + 6x + 4$

    H. $x^2 + 6x + 10$

    J. $3x^2 + 18x + 10$

    K. $3x^2 - 18x + 4$

23. $\left(a - \dfrac{1}{3}b\right)^2 = ?$

    A. $a^2 - \dfrac{1}{9}b^2$

    B. $a^2 - \dfrac{2}{3}ab + \dfrac{1}{9}b^2$

    C. $a^2 - \dfrac{2}{3}ab + \dfrac{1}{3}b^2$

    D. $a^2 + b^2$

    E. $a^2 - \dfrac{1}{3}ab - \dfrac{1}{9}b^2$

4. When $x - y = 8$, what is the value of $3\,(x - y) + (x - y)^2 - \dfrac{x - y}{4} - 15?$

    F. 15

    G. 55

    H. 71

    J. 72

    K. 101

Common Algebra Mistakes
Algebra and Arithmetic Problem Set 3

25. If $x = 3a + 5$ and $y = 6 - a$, which of the following expresses x in terms of y?

   A. $x = 3y - 13$

   B. $x = \dfrac{y - 11}{3}$

   C. $x = \dfrac{y - 23}{3}$

   D. $x = 3y + 15$

   E. $x = 23 - 3y$

26. If $f(x) = 2x^2 + 18$ and $f(2a) = 34$, what is a possible value of a?

   F. 0

   G. 1

   H. $\sqrt{2}$

   J. 2

   K. 3

27. The graph below shows the distribution of Jane's $5,000 business expenses. The amount Jane paid for office supplies was shared equally with 4 other clients. What was the total cost of Jane's portion of the office supplies?

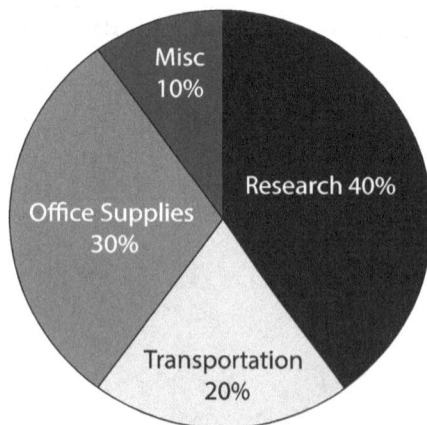

   A. $1,250

   B. $1,500

   C. $5,000

   D. $6,000

   E. $7,500

Common Algebra Mistakes
Algebra and Arithmetic Problem Set 3

28. The square of x is equal to 16 times the square of y. If x is one more than 4 times y, what is the value of x?

F. $-\dfrac{1}{4}$

G. $-\dfrac{1}{8}$

H. 0

J. $\dfrac{1}{3}$

K. $\dfrac{1}{2}$

29. The result when three less than negative four times a number is subtracted from the same number squared is zero. What is one possible value of the unknown number?

A. -1

B. $-\sqrt{3}$

C. 1

D. 3

E. 2

DO YOUR FIGURING HERE

Common Algebra Mistakes
Algebra and Arithmetic Problem Set 3

Answer Key

#	Answer	Frequency	Difficulty
1	D	popular	3
2	H	popular	2
3	C	popular	3
4	G	popular	2
5	E	popular	2
6	K	popular	1
7	D	popular	1
8	J	popular	2
9	E	popular	2
10	K	popular	2
11	B	popular	2
12	G	popular	2
13	A	popular	2
14	G	popular	1
15	A	popular	2
16	H	popular	2
17	B	popular	2
18	F	popular	1
19	E	popular	2
20	F	popular	3
21	A	popular	1
22	H	popular	1
23	B	popular	2
24	H	popular	2
25	E	popular	3
26	H	popular	3
27	B	popular	3
28	K	popular	4
29	A	popular	3

Algebraic Manipulation and Patterns
Algebra and Arithmetic Problem Set 4

1. If $13b = 22$, what is the value of $35 - 13b$?
   A. 7
   B. 10
   C. 13
   D. 16
   E. 22

DO YOUR FIGURING HERE

2. If $x^3 + r = 1(x^3 + 5)$, what is the value of r?
   F. 1
   G. 2
   H. 3
   J. 4
   K. 5

3. If $3cd = 12$ then $(cd)^2 =$
   A. 8
   B. 9
   C. 16
   D. 64
   E. 144

4. If $ab = 7$, what is the value of $\dfrac{21a^2}{a^3b}$?
   F. 3
   G. 5
   H. 7
   J. 14
   K. 21

5. If $n^2 - 2mn + m^2 = mn$, then which of the following is equivalent to $n^2 + m^2$?
   A. $-mn$
   B. $-2mn$
   C. mn
   D. $3mn$
   E. $(mn)^2$

Algebraic Manipulation and Patterns
Algebra and Arithmetic Problem Set 4

6. If $m + n = 6$ and $c + d = 7$, what is the value of $mc + nd + md + nc$?

    F. 21

    G. 36

    H. 42

    J. 64

    K. 84

7. If $a^2 = b^2 + 100$, which of the following expressions must equal 100?

    A. $2(a - b)$

    B. $(a + b)^2$

    C. $a^2 - b^2$

    D. $a^2 + b^2$

    E. $(a - b)^2$

8. If $y = 3x - 1$ and $x > 4$, which of the following represents all the possible values for y?

    F. $y > 11$

    G. $y < 11$

    H. $y > 13$

    J. $y < 13$

    K. $11 < y < 13$

9. If $a^2 - b^2 = 20$ and $a - b = 4$, what is the value of a?

10. If $s^2 + t^2 = 2st$, then s must equal

    F. 0

    G. 1

    H. t

    J. $2t$

    K. $2st$

DO YOUR FIGURING HERE

2019-1

Algebraic Manipulation and Patterns
Algebra and Arithmetic Problem Set 4

11. If $a^n = 12$ and $a^s = 3$, what does a^{n-s} equal?

  A. 4

  B. 9

  C. 10

  D. 15

  E. 36

DO YOUR FIGURING HERE

12. If $3a^2 = 12$ and $3b^2 = 21$, what is the value of $a^2 b^2$?

  F. -28

  G. -9

  H. 0

  J. 9

  K. 28

13. If $\dfrac{x}{5} + \dfrac{y}{5} = 25$, what is the value of $x + y$?

  A. 5

  B. 10

  C. 25

  D. 50

  E. 125

4. If $(x - 3)(x + 3) = 16$, what is the value of x^2?

  F. 5

  G. 10

  H. 25

  J. 50

  K. 125

Algebraic Manipulation and Patterns
Algebra and Arithmetic Problem Set 4

15. If $a^2 - b^2 = 63$ and $a - b = 7$, what is the value of $a + b$?

   A. 9

   B. 10

   C. 11

   D. 12

   E. 13

DO YOUR FIGURING HERE

16. If a and b are positive numbers, then the inequality $a\sqrt{7} > b\sqrt{2}$ is equivalent to which of the following?

   F. $a < 15b$

   G. $a < \dfrac{2}{7}b$

   H. $a > \dfrac{2}{7}b$

   J. $a^2 < \dfrac{2}{7}b^2$

   K. $a^2 > \dfrac{2}{7}b^2$

17. What is the value of $\dfrac{x(x + y - x - y)y}{2xy}$?

   A. 0

   B. $\dfrac{1}{2}$

   C. 4

   D. $\dfrac{5}{5}$

   E. 6

18. If $x^2 + y^2 = 40$ and $xy = 16$, what is the value of $(x + y)^2$?

   F. 49

   G. 58

   H. 64

   J. 72

   K. 81

Algebraic Manipulation and Patterns
Algebra and Arithmetic Problem Set 4

19. What is the value of $\dfrac{3a(a^2 - b)}{a^3 - ab}$?

  A. -2

  B. 0

  C. 1

  D. 3

  E. 9

20. If $(x + y)^2 = 100$ and $x^2 + y^2 = 68$, what is the value of $2xy$?

  F. -54

  G. -32

  H. 0

  J. 32

  K. 54

21.
$$c = d^2$$
$$c = \sqrt{e}$$
$$c = \frac{125}{d}$$

In the system of equations above, if $d > 1$, what is the value of e?

22. If $a^2 = 10$ and $b^2 = 25$, then $(2a + b)^2$ could equal which of the following?

  F. 35

  G. 225

  H. $65 - 4\sqrt{10}$

  J. $65 + 20\sqrt{10}$

  K. $65 + 10\sqrt{5}$

Algebraic Manipulation and Patterns
Algebra and Arithmetic Problem Set 4

23.
$$\frac{6}{\sqrt{x+3}} = 5$$

For $x > -3$, which of the following equations is equivalent to the equation above?

A. $6 = 5(x+3)$

B. $6 = 25(x+3)$

C. $36 = 25(x+2)$

D. $36 = 25(x+3)$

E. $36 = 5(x+3)$

24. Each of the following is equivalent to $\frac{s}{t}(v+tu)$ EXCEPT:

F. $\frac{sv}{t} + su$

G. $\frac{sv + stu}{t}$

H. $\frac{s(v+tu)}{t}$

J. $\frac{sv}{t} + \frac{stu^2}{tu}$

K. $\frac{sv}{t} + stu$

25. If $3^{2y} + 3^{2y} + 3^{2y} + 3^{2y} = 4(3^{y+5})$, what is the value of y?

26. Which of the following is NOT equivalent to $\frac{d^2 - 64}{16}$?

F. $\frac{1}{16}(d+8)(d-8)$

G. $\frac{d^2}{16} - 4$

H. $\frac{d^2}{16} - 64$

J. $\left(\frac{d}{4}\right)^2 - 2^2$

K. $\frac{(d-8)(d+8)}{16}$

Algebraic Manipulation and Patterns
Algebra and Arithmetic Problem Set 4

27. If $a = 2 + \dfrac{2}{b}$ and $b > 2$, then a could equal

A. $\dfrac{1}{2}$

B. 1

C. $\dfrac{3}{2}$

D. 2

E. $\dfrac{5}{2}$

DO YOUR FIGURING HERE

28. If $a^2 + b^2 = 45$ and $ab = 18$, which of the following equals $(a + b)^2$?

F. 30

G. 49

H. 64

J. 81

K. 100

29. If $(x + y)(3x - 3y) = 27$, what is the value of $3x^2 - 3y^2$?

A. 16

B. 21

C. 27

D. 29

E. 33

30. If $x^2 + 2xy + y^2 = 144$, what is one possible value of $x + y$?

F. -18

G. -12

H. 6

J. 9

K. 18

31. If $t - z < -z$, which of the following must be true?

    A. $2z < 0$

    B. $z < 0$

    C. $z > 0$

    D. $t > 0$

    E. $t < 0$

DO YOUR FIGURING HERE

32.
$$a + b + c = 500$$
$$a + b + 5c = 1100$$

In the system of equations above, what is the value of $a + b$?

33. If $x + y = 4$, what is the value of $x^2 + 2xy + y^2$?

    A. 10

    B. 16

    C. 21

    D. 25

    E. 36

34. If $ab = 21$ and $a - b = 7$, what is the value of $a^2b - ab^2$?

    F. 14

    G. 147

    H. 217

    J. 256

    K. 284

35. If $abcd = -1$ and $acde = 0$, which of the following must be true?

    A. $c = 0$

    B. $e = 0$

    C. $a > 1$

    D. $a < 0$

    E. $abc < 1$

Algebraic Manipulation and Patterns
Algebra and Arithmetic Problem Set 4

36. If $(x + y)^2 = 169$ and $(x - y)^2 = 9$, what is the value of xy?

   F. 3

   G. 18

   H. 40

   J. 80

   K. 160

DO YOUR FIGURING HERE

37. If $\dfrac{(m + n)}{y - x} = \dfrac{4}{5}$, then $\dfrac{(7m + 7n)}{(4y - 4x)} =$

   A. $\dfrac{4}{7}$

   B. $\dfrac{5}{7}$

   C. $\dfrac{7}{5}$

   D. $\dfrac{7}{4}$

   E. $\dfrac{28}{5}$

38.
$$a^5 b^4 c^3 > a^6 b^3 c^3$$
If a, b, and c are positive and the above statement is true, which of the following must be true?

       I. $b > a$

      II. $b > c$

     III. $a > c$

   F. I only

   G. II only

   H. III only

   J. II and III only

   K. I, II and III

39. If a and b are positive numbers, a is greater than 1, and $b = 2a - \dfrac{1}{a}$, which of the following must be true?

$$\text{I. } b > 1$$
$$\text{II. } a = b$$
$$\text{III. } ab < b^2$$

A. I only

B. II only

C. III only

D. I and III only

E. I, II and III

DO YOUR FIGURING HERE

40. If $s > 0$, $m^2 + n^2 = s$ and $mn = s + 5$, what is $(m + n)^2$ in terms of s?

F. $s + 10$

G. $3s + 5$

H. $3s + 10$

J. $3s + 20$

K. $3s + 30$

41.
$$50a + 77b = 2077$$
If a and b are positive integers in the equation above, what is one possible value of $a + b$?

42. If x and y are positive numbers and $\dfrac{2x}{y} > 2xy$, which of the following must be true?

$$\text{I. } y < 1$$
$$\text{II. } x > 1$$
$$\text{III. } x > y$$

F. I only

G. II only

H. III only

J. I and II

K. II and III

Algebraic Manipulation and Patterns
Algebra and Arithmetic Problem Set 4

43. If $xy \neq x$ and $y = \dfrac{1}{x}$, which of the following expressions is equivalent to $\dfrac{(1 - x)}{(1 - y)}$?

  A. $\dfrac{-1}{y}$

  B. $-y$

  C. x^2

  D. 0

  E. y

44.
$$ab = a + b$$
If a and b are positive numbers and $b > 2$, what are all possible values of a that satisfy the equation above?

  F. $a > 0$

  G. $a < 0$

  H. $0 < a < 1$

  J. $-1 < a < 0$

  K. $1 < a < 2$

45. If $0 \leq a \leq 2b$ and $(a + 2b)^2 - (a - 2b)^2 \geq 144$, what is the least possible value of b?

46.
$$a = 4b$$
$$b = 10c$$
$$a = xc$$
For the system of equations above, if $a \neq 0$, what is the value of x?

Algebraic Manipulation and Patterns
Algebra and Arithmetic Problem Set 4

Answer Key

#	Answer	Frequency	Difficulty
1	C	popular	1
2	K	popular	2
3	C	popular	1
4	F	popular	2
5	D	popular	2
6	H	popular	2
7	C	popular	2
8	F	popular	3
9	4.5 or $\frac{9}{2}$	popular	3
10	H	popular	4
11	A	popular	2
12	K	popular	3
13	E	popular	3
14	H	popular	2
15	A	popular	3
16	K	popular	3
17	A	popular	1
18	J	popular	3
19	D	popular	3
20	J	popular	3
21	625	popular	3
22	J	popular	3
23	D	popular	2
24	K	popular	3
25	5	popular	3
26	H	popular	3
27	E	popular	3
28	J	popular	3
29	C	popular	2
30	G	popular	3
31	E	popular	2
32	350	popular	1
33	B	popular	3
34	G	popular	3
35	B	popular	1
36	H	popular	2
37	C	popular	3
38	F	popular	2
39	D	popular	2
40	H	popular	3
41	41	popular	4
42	F	popular	3
43	A	popular	4
44	K	popular	4
45	3	popular	4
46	40	popular	3

Translating English to Algebra
Algebra and Arithmetic Problem Set 5

1. When twice a certain number is decreased by 10, the result is 7. What is the number?

DO YOUR FIGURING HERE

2. If 7 more than 3 times a number is equal to 16, what is 5 times the number?

   F. 3

   G. 15

   H. 20

   J. 22.5

   K. 30

3. Trucks E, F, and G all passed through a checkpoint during a race traveling at different speeds. Truck E's speed was 4 times Truck F's speed, and Truck G's speed was twice Truck E's speed. What was Truck G's speed, in terms of miles per hour, if Truck F's speed was 15 miles per hour?

   A. 30

   B. 45

   C. 60

   D. 90

   E. 120

4. If $\frac{5}{8}$ of x is 120, what is $\frac{3}{8}$ of x?

   F. 36

   G. 72

   H. 108

   J. 192

   K. 576

5. The smaller of two numbers is 5 less than half of the larger. The sum of twice the larger number and 3 times the smaller number is 89. If x is the larger number, which equation below determines the correct value?

A. $3x + 2(2x - 10) = 89$

B. $3x + 2\left(\dfrac{x}{2} + 5\right) = 89$

C. $3x + (x + 5) = 89$

D. $2x + 3(2x - 10) = 89$

E. $2x + 3\left(\dfrac{x}{2} - 5\right) = 89$

DO YOUR FIGURING HERE

6. The price of one slice of pizza and a side salad is $4.50. The cost of two slices of pizza and a side salad together $6.25. What is the cost of a side salad?

F. $1.25

G. $1.75

H. $2.25

J. $2.75

K. $3.75

7. If 6 times a number a is added to 24 the sum is positive. Which of the following gives the possible value(s) for a?

A. 0 only

B. 4 only

C. 18 only

D. All $a > \text{-}4$

E. All $a < \text{-}4$

Translating English to Algebra
Algebra and Arithmetic Problem Set 5

8. Before cooking a meal, Joel spends 10 minutes gathering ingredients, double checking the recipe, and laying everything out that he may need while cooking. The equation $t = 5n + 10$ models the time, t minutes, that Joel budgets for cooking a meal with n ingredients. Which of the following statements is necessarily true according to Joel's model?

   F. He budgets 10 minutes for ingredients he needs to chop and 5 minutes for those that don't need to be chopped.

   G. He budgets a 5 minute break between ingredients, to let the ingredient cook.

   H. He budgets 15 minutes per ingredient.

   J. He budgets 10 minutes per ingredient.

   K. He budgets 5 minutes per ingredient.

9. Honey is planning to make a flat patio in her back yard using bricks that are 4 inches wide by 8 inches long. What is the minimum number of bricks she will need if she wants her patio to cover a rectangular area of 10 feet by 12 feet and all of the paving blocks will face the same way?

   A. 48

   B. 250

   C. 360

   D. 375

   E. 540

10. A large rectangular prism has sides that are three times longer than those of a similar small rectangular prism. The volume of the large rectangular prism is how many times the volume of the small rectangular prism?

   F. 3

   G. 6

   H. 9

   J. 27

   K. 81

Translating English to Algebra
Algebra and Arithmetic Problem Set 5

11. On January 1st a winter coat was $120. On February 1st, the price was reduced by 15%. On March 1st, the price was further reduced by 30% of the February 1st price and marked "Final Sale". What percent of the original price was the "Final Sale" price?

   A. 40.5%

   B. 45%

   C. 55%

   D. 56%

   E. 59.5%

12. Survey Inc. was hired to determine the average age of a group of 500 people. The average age of the 300 males was 36 and the average age of the 200 females was 42. What is the approximate age of the 500 people who were surveyed?

   F. 37

   G. 38

   H. 39

   J. 40

   K. 41

13. In the 1916 United States Presidential election, there were a total of 531 electoral votes cast, and each vote went to Woodrow Wilson or Charles Evans Hughes. Wilson received 23 more electoral votes than Hughes. Approximately what percent of the 531 electoral votes were cast for Wilson?

   A. 49.7

   B. 50

   C. 51.6

   D. 52.2

   E. 53.5

DO YOUR FIGURING HERE

Translating English to Algebra
Algebra and Arithmetic Problem Set 5

14. Jane ate t tacos on Sunday, 4 times as many tacos on Monday as on Sunday, and 9 more tacos on Tuesday than on Sunday. What is the average number of tacos that Jane ate each day?

    F. $6t + 9$

    G. $\left(\dfrac{5t}{3}\right) - 3$

    H. $\left(\dfrac{5t}{3}\right) + 3$

    J. $2t - 3$

    K. $2t + 3$

15. A farmer can raise s sheep for every acre of land. If at the end of the season she sells each sheep for $200, how many dollars can she earn on r acres of land?

    A. $200sr$

    B. $\dfrac{200sr}{5}$

    C. $\dfrac{200s}{5}$

    D. $\dfrac{s}{r}$

    E. $\dfrac{200sr}{200r}$

16. The depth of water in a bathtub is 38 inches and is being drained by 1 inch per minute. The depth of water in a second bathtub is 32 inches and is being drained by $\dfrac{1}{2}$ inches per minute. If the depths of both bathtubs continue to be drained at these constant rates, in how many minutes will the bathtubs have the same depth?

    F. 3

    G. 4

    H. 12

    J. 36

    K. 140

DO YOUR FIGURING HERE

17. During an art sale, a customer can buy one print for a dollars. Each additional print the customer buys costs b dollars less than the first print. For example, the cost of the second print is $a - b$ dollars. Which of the following represents the customer's cost, in dollars, for p prints bought during the sale?

   A. $a + p(a - b)$

   B. $a + pab$

   C. $a + (p - 1)(a - b)$

   D. $\dfrac{a + p(a - b)}{b}$

   E. $a + p + (a - b)$

18. A car rental company rents their cars for $15 per day with an additional $0.40 for every mile driven after the first 20 miles. Which of the following expressions represents the cost, in dollars, of renting a van for 1 day and driving it m miles where $m > 20$?

   F. $15 + 40m$

   G. $40 + 15m$

   H. $55m$

   J. $15 + .4m$

   K. $15 + .4(m - 20)$

19. The cost of a telephone call using Talk Too Much telephone company is $1.00 for any time up to and including 15 minutes and $0.10 per minute thereafter. The cost using the Phone Me Telephone Company is $0.08 per minute for any amount of time. For a call that lasts m minutes, the cost is the same to use both carriers. If m is an integer greater than 15, what is the value of m?

Translating English to Algebra
Algebra and Arithmetic Problem Set 5

0. A cellular phone carrier charges its customers $15 per month for up to 500 text messages and $0.30 per text over 500 in that month. Which of the following expressions gives a customer's total monthly text message charges, in dollars, for sending t text messages, where $t > 500$?

F. $15.3t$

G. $15 + 150t$

H. $15 + .3t$

J. $15 + .3(t - 500)$

K. $515 + .3t$

DO YOUR FIGURING HERE

Translating English to Algebra
Algebra and Arithmetic Problem Set 5

Answer Key

#	Answer	Frequency	Difficulty
1	$\frac{17}{2}$ or 8.5	popular	1
2	G	popular	1
3	E	popular	2
4	G	popular	2
5	E	popular	2
6	J	popular	2
7	D	popular	2
8	K	popular	2
9	E	popular	2
10	J	rare	3
11	E	popular	3
12	G	average	3
13	D	popular	3
14	K	popular	3
15	A	popular	3
16	H	popular	3
17	C	average	4
18	K	average	4
19	25	average	3
20	J	average	4

Testing Values
Algebra and Arithmetic Problem Set 6

. If $3a - 2 = b$, which of the following must be equal to $9a - 6$?

A. $b + 5$

B. $b + b$

C. $2b$

D. $2b + 5$

E. $3b$

. If $4a = 9b$ and $6b = 8c$, what does a equal in terms of c?

F. $\dfrac{1}{3}c$

G. $\dfrac{32}{81}c$

H. $3c$

J. $10c$

K. $20c$

. If $3x < y < -1$, which of the following must be the greatest?

A. $-(x - y)^2$

B. $-(x + y)$

C. $-(x - y)$

D. $x + y$

E. 0

. If a is an odd integer and b is an even integer, which of the following must be an odd integer?

$$\text{I. } (a + 1)\, b$$
$$\text{II. } (b + 1)\, a$$
$$\text{III. } (a + 1)\, \text{-}b$$

F. I only

G. II only

H. III only

J. I and II only

K. I and III only

5. If $0 < a < b < c < 3$, which of the following could not be more than 3?

   A. $\dfrac{a}{c}$

   B. $\dfrac{c}{a}$

   C. $3a$

   D. $3b$

   E. b^2

DO YOUR FIGURING HERE

6. If $n(x - 3) + 4 = m$, what is $(x - 3)$ in terms of m and n?

   F. $\dfrac{m - n}{4}$

   G. $\dfrac{m + 4}{n}$

   H. $\dfrac{m - 4}{n}$

   J. $\dfrac{m}{n} - 4$

   K. $\dfrac{m}{n} + 4$

7. If B is the true midpoint of AC, which of the following must be true?

$$\text{I. } 2BC = AC$$
$$\text{II. } 0.5AC = AB$$
$$\text{III. } 2AC = 4AB$$

   A. I only

   B. II only

   C. I and II only

   D. I and III only

   E. I, II, and III

Testing Values
Algebra and Arithmetic Problem Set 6

8. $a, b,$ and c are positive integers and $b > 1$. If $ab = 21$ and $bc = 35$, which of the following must be true?

F. $b > c > a$

G. $c > b > a$

H. $a > b > c$

J. $a > c > b$

K. $b > a > c$

DO YOUR FIGURING HERE

9. Bill will be b years old in 4 years. How old was Bill, in terms of b, 3 years ago?

A. $b - 4$

B. $b - 5$

C. $b - 6$

D. $b - 7$

E. $b - 8$

10. If $a < b < 0 < c$, which of the following could be true?

$$\text{I. } ab > c$$
$$\text{II. } ac > b$$
$$\text{III. } bc = -1$$

F. none

G. I only

H. I and II only

J. I and III only

K. I, II, and III

1. If $3u = 5w + 6$, what is the value of w in terms of u?

A. $\dfrac{(3u - 6)}{5}$

B. $\dfrac{(6 - 3u)}{5}$

C. $\dfrac{3(u - 6)}{6}$

D. $\dfrac{(3u - 5)}{6}$

E. $15u$

12. If $a > 0$ and $3a = 4b = 5c = 6d$, which of the following is true?

   F. $a < c < b < d$

   G. $a < b < c < d$

   H. $d < c < b < a$

   J. $d < b < a < c$

   K. $d < c < a < b$

DO YOUR FIGURING HERE

13. $3a = \dfrac{3b}{3}$. If a and b are positive integers, what is the value of b in terms of a?

   A. $3a$

   B. $6a$

   C. $9a$

   D. $a + 3$

   E. $a + 4$

14. The price of a television set was first increased by 15%. Then the new price was decreased by 20%. The final price was what percent of the initial price?

   F. 80%

   G. 82%

   H. 92%

   J. 95%

   K. 102%

15. Amanda has a apples, b bananas, and k kiwis. She has 4 more apples than bananas and 7 fewer kiwis than bananas. If a, b, and k are positive integers, what is the value of k in terms of a?

   A. $a - 11$

   B. $a - 3$

   C. $a + 1$

   D. $a + 3$

   E. $2a$

Testing Values
Algebra and Arithmetic Problem Set 6

16. If $x^3 = -y$ for all real numbers x and y, which of the following could be true?

$$\text{I. } y < 0$$
$$\text{II. } y = 0$$
$$\text{III. } y > 0$$

F. I only

G. II only

H. III only

J. I and III only

K. I, II, and III

17. If an integer n is divisible by 2, 3, 8, and 12, then what will be the next largest integer divisible by these numbers?

A. $n + 8$

B. $n + 12$

C. $n + 24$

D. $n + 36$

E. $n + 44$

18. There are l liters of soda to serve at a party. After k liters have been consumed, then, in terms of l and k, what percent of the soda has NOT been consumed?

F. $\dfrac{100k}{l}\%$

G. $\dfrac{100l}{k}\%$

H. $\dfrac{100(l - k)}{l}\%$

J. $\dfrac{k}{100(l - k)}$

K. $\dfrac{100k}{l}$

Testing Values
Algebra and Arithmetic Problem Set 6

19. John divides his new book club novel into t sections of equal pages to help him make a timetable for completing the novel. If $t > 8$ and he plans to read 2 sections a night until the book is finished, what percentage of the novel will be left to read after three nights?

   A. $\dfrac{(t+3)}{100t}$

   B. $\dfrac{100\,(t\ -\ 3)}{t}$

   C. $\dfrac{100\,(t\ -\ 6)}{t}$

   D. $\dfrac{(t+6)}{100}$

   E. $\dfrac{100t}{t\ -\ 6}$

20. An aquarium needs to feed its sharks fresh fish everyday. A white tipped reef shark eats b sea bass and f flounder each day. A black tipped reef shark eats $2b$ sea bass and c cod each day. If there are s white tipped sharks, and $2s$ blacked tipped sharks, how many fresh fish does the aquarium need every day?

   F. $2sf + 2sb + 2sc$

   G. $2sc + 5sb + sf$

   H. $2sb + 6sf + sc$

   J. $4sf + 3sc + 2sb$

   K. $3sc + 2sf + 4sb$

Testing Values
Algebra and Arithmetic Problem Set 6

1. Raquel's fishing boat caught p pounds of tuna fish last week. Raquel gave 20 percent of the fish to her crew members and then packed the rest in tin cans for storage. If each can holds r ounces, how many tin cans can Raquel pack? (16 ounces = 1 pound)

   A. $\dfrac{2p}{5r}$

   B. $\dfrac{64p}{5r}$

   C. $\dfrac{4pr}{5}$

   D. $\dfrac{65p}{5r}$

   E. $\dfrac{64r}{p}$

DO YOUR FIGURING HERE

Testing Values
Algebra and Arithmetic Problem Set 6

Answer Key

#	Answer	Frequency	Difficulty
1	E	popular	1
2	H	popular	2
3	B	popular	2
4	G	popular	2
5	A	popular	2
6	H	popular	2
7	E	popular	2
8	F	popular	2
9	D	popular	2
10	K	average	2
11	A	popular	2
12	H	popular	2
13	A	popular	1
14	H	popular	3
15	A	popular	4
16	K	average	3
17	C	popular	2
18	H	popular	2
19	C	popular	4
20	G	popular	5
21	B	popular	5

2019-1

Least Common Multiple and Greatest Common Factor
Algebra and Arithmetic Problem Set 7

1. What is the least common multiple of 3, 12, and 15?

   A. 3

   B. 9

   C. 60

   D. 180

   E. 540

2. What is the greatest common factor of 132, 45, and 144?

   F. 3

   G. 5

   H. 9

   J. 12

   K. 45

3. The least common multiple of 2 numbers is 63. The greater of the 2 numbers is 21; what is the maximum value of the 2nd number?

   A. 3

   B. 6

   C. 9

   D. 14

   E. 21

4. What is the least common multiple of $x, 3y^2, 5xy$, and $6x^3$?

   F. $3x^y$

   G. x^3y^2

   H. $15x^2y$

   J. $15x^3y^2$

   K. $30x^3y^2$

DO YOUR FIGURING HERE

Least Common Multiple and Greatest Common Factor
Algebra and Arithmetic Problem Set 7

5. What is the least commmon denominator when adding the fractions $\frac{4}{30}$, $\frac{5}{36}$, and $\frac{6}{45}$?

    A. 120

    B. 140

    C. 180

    D. 1600

    E. 4860

DO YOUR FIGURING HERE

6. What is the greatest common factor of the following 3 monomials? $3a^2b, 17a^5b^3, 51a^3b^2$

    F. a^2b

    G. $3a$

    H. $3a^2b$

    J. $51a^3b^2$

    K. $51a^5b^3$

7. What is the least common multiple of the numbers $3, 5, 7, 15$, and 35?

    A. 75

    B. 90

    C. 105

    D. 115

    E. 55, 125

8. What is the least common multiple of $40, 50$, and 60?

    F. 200

    G. 400

    H. 500

    J. 600

    K. 900

2019-1

Least Common Multiple and Greatest Common Factor
Algebra and Arithmetic Problem Set 7

9. What is the greatest common factor of $120x^3y^2, 300xy^2,$ and $75x^5y^3$?

  A. $5xy^2$

  B. $15xy^2$

  C. $15x^2y^2$

  D. $30x^2y^2$

  E. $30x^2y^3$

10. The least common multiple (LCM) of 2 numbers is 360. The greater number is 40. What is the maximum value of the other number?

  F. 12

  G. 18

  H. 20

  J. 24

  K. 36

11. Which of the following is the least common denominator for the expression below?
$$\frac{1}{2^2 \cdot 3^3 \cdot 4^4} + \frac{1}{3^4 \cdot 4^2} + \frac{1}{2^3 \cdot 3^2 \cdot 4^3}$$

  A. $2 \cdot 3 \cdot 4$

  B. $2^2 \cdot 3^2 \cdot 4^3$

  C. $2^3 \cdot 3^4 \cdot 4^4$

  D. $2^3 \cdot 3^3 \cdot 4^3$

  E. $2^3 \cdot 3^3 \cdot 4^4$

12. What is the least common denominator of the fractions $\frac{5}{7}, \frac{3}{56}, \frac{7}{5}$?

  F. 56

  G. 112

  H. 140

  J. 280

  K. 560

13. If a and b are positive integers such that the greatest common factor of x^4y and x^2y^3 is 175, then which of the following could y equal?

    A. 5

    B. 7

    C. 9

    D. 25

    E. 35

DO YOUR FIGURING HERE

Least Common Multiple and Greatest Common Factor
Algebra and Arithmetic Problem Set 7

Answer Key

#	Answer	Frequency	Difficulty
1	C	popular	1
2	F	popular	1
3	C	popular	2
4	K	popular	2
5	C	popular	2
6	F	popular	2
7	C	popular	2
8	J	popular	2
9	B	popular	2
10	K	popular	2
11	C	popular	2
12	J	popular	2
13	B	popular	3

Proportions
Algebra and Arithmetic Problem Set 8

1. A machine can manufacture 36 pens in one hour. At this rate, how many pens are manufactured in 5 minutes?

   A. One

   B. Three

   C. Four

   D. Five

   E. Six

DO YOUR FIGURING HERE

2. During heavy traffic, a person could bike 3 times as fast as a car. If a car was traveling at 5 miles per hour during heavy traffic, how many hours would it take for a cyclist to bike 20 miles?

   F. 1 hour

   G. $1\frac{1}{3}$ hours

   H. $1\frac{1}{2}$ hours

   J. 2 hours

   K. 15 hours

3. On Saturday, Jim rode his bike 8 miles in 30 minutes. If he rode at the same pace for 45 minutes on Sunday, how far did he ride on Sunday?

   A. $5\frac{1}{3}$ miles

   B. $10\frac{1}{3}$ miles

   C. 12 miles

   D. 14 miles

   E. 16 miles

Proportions
Algebra and Arithmetic Problem Set 8

4. According to the circle graph below, how many of the pizza toppings individually represent more than 25 percent of the total sales?

Pizza Sales by Topping

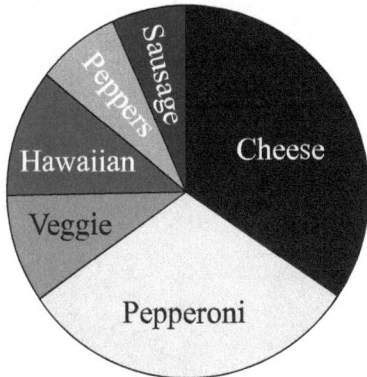

F. Zero

G. One

H. Two

J. Three

K. Four

. At Central High School, 75 of the 120 juniors are members of one of the athletic teams. The fraction of the number of seniors who are members of an athletic team is equal to the fraction of the number of juniors who are members of an athletic team. How many of the 136 seniors are members of an athletic team?

A. 75

B. 78

C. 80

D. 85

E. 90

DO YOUR FIGURING HERE

Proportions
Algebra and Arithmetic Problem Set 8

6. Pedro's new bike has wheels that are 26 inches in diameter. Pedro rode his bike so that it traveled along a straight path for 30 revolutions. How far does each wheel travel from its starting point?

F. 390

G. 780

H. 390π

J. 676π

K. 780π

DO YOUR FIGURING HERE

7. How many minutes would it take a car to travel 60 miles at a constant speed of 100 miles per hour?

A. 24

B. 36

C. 60

D. 96

E. 100

8. The circle below has a radius of 6 inches. If $\angle NOP$ has a measure of 100°, what is the area of the shaded sector?

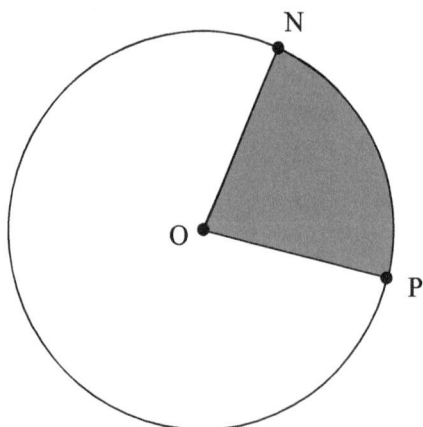

F. 3.6π

G. 5π

H. 6π

J. 10π

K. 12π

Proportions
Algebra and Arithmetic Problem Set 8

9. Ms. Larkin spends 20 minutes of every hour-long class reviewing homework and teaches 5 classes every day. How many minutes does she spend reviewing homework every day?

   A. 50 minutes

   B. 60 minutes

   C. 80 minutes

   D. 100 minutes

   E. 120 minutes

10. Jordan is making posters for the school election. It takes her one hour to make 6 small posters and it takes her an hour and a half to make 6 large posters. How many more small posters than large posters can she make in 3 hours?

   F. 3

   G. 6

   H. 9

   J. 12

   K. 18

11. There are 0.621 miles in 1 kilometer. If a car travels at 25 mph, how many kilometers is it moving per minute?

12. There are 3.79 liters in 1 gallon. If a generator uses 14 gallons of fuel in a day, how many liters does it use in 4 weeks?

DO YOUR FIGURING HERE

13. In the figure below, $\triangle ABC$ is similar to $\triangle DEF$. What is the value of x?

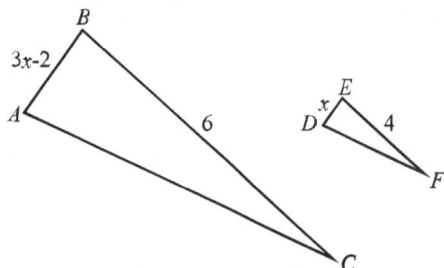

DO YOUR FIGURING HERE

A. $\dfrac{1}{3}$

B. $\dfrac{3}{4}$

C. $\dfrac{6}{7}$

D. 1

E. $\dfrac{4}{3}$

14. Cassidy is making a map of her route from her home, A, to her school, B, and her friend's home, C. She shows the 2-mile distance between home and school as 13 centimeters on the map. The distance between Cassidy's school and her friend's home is 1.5 miles. How long should the line between her school and her friend's home be on the map? Round to the nearest hundredth of a centimeter.

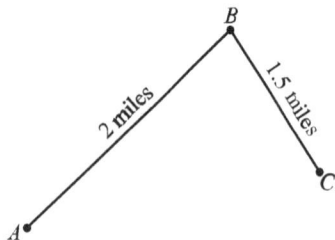

F. 5.34 cm

G. 7.89 cm

H. 8.67 cm

J. 9.75 cm

K. 17.33 cm

Proportions
Algebra and Arithmetic Problem Set 8

15. In the figure below, $\triangle ABC$ is similar to $\triangle DEF$. The area of $\triangle ABC$ is 36. What is the area of $\triangle DEF$? Round to the nearest integer.

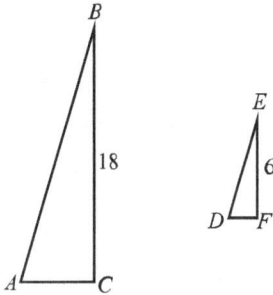

A. 4

B. 8

C. 10

D. 12

E. 18

DO YOUR FIGURING HERE

16. There are 0.621 miles in 1 kilometer. How many kilometers is a 26.2-mile marathon?

17. During a leap year, February has 29 days. How many seconds are there in February if it is a leap year?

18. Janet can paint 3 square inches of her wall per second. How many hours will it take her to paint a wall with the dimensions 8 feet by 15 feet?

19. On Tuesday, Restaurant Good Food served 300 total customers. Of these, 210 people ordered cheeseburgers. On Saturday, the restaurant served 280 cheeseburgers. If they served the same proportion of cheeseburgers, then how many customers did they serve on Saturday?

A. 280

B. 295

C. 310

D. 370

E. 400

20. If $\dfrac{x-3y}{x+2y} = \dfrac{4}{9}$, then $\dfrac{x}{y} = ?$

    F. $\dfrac{1}{7}$

    G. $\dfrac{4}{9}$

    H. 4

    J. 7

    K. 12

DO YOUR FIGURING HERE

Proportions
Algebra and Arithmetic Problem Set 8

Answer Key

#	Answer	Frequency	Difficulty
1	B	popular	2
2	G	popular	1
3	C	popular	2
4	H	popular	2
5	D	popular	2
6	K	popular	2
7	B	popular	2
8	J	average	2
9	D	popular	1
10	G	popular	2
11	0.671 km/ min	popular	2
12	1485.68 liters	popular	2
13	E	average	2
14	J	average	2
15	A	average	2
16	$x = 42.2$ km	popular	2
17	$2,505,600$ seconds	popular	2
18	1.6 hours	popular	2
19	E	popular	3
20	J	popular	3

Ratios
Algebra and Arithmetic Problem Set 9

1. If the ratio of blue pens to black pens in a drawer is 2 to 7 and the total number of blue and black pens is 99, then how many black pens are there in the drawer?

   A. 11

   B. 22

   C. 55

   D. 77

   E. 88

2. The ratio of boys to girls in gym class is 3 to 5. Which of the following could be the total number of students in the gym class?

   F. 6

   G. 13

   H. 15

   J. 23

   K. 24

3. The ratio of apples to oranges in a basket is $\frac{1}{2}$. If there are 24 oranges in the basket, how many apples are there?

   A. 8

   B. 12

   C. 24

   D. 36

   E. 48

4. If the ratio of x to y is 6 to 5, then what is the ratio of x to $2y$?

   F. 2 to 5

   G. 3 to 5

   H. 9 to 16

   J. 12 to 5

   K. 27 to 10

DO YOUR FIGURING HERE

Ratios
Algebra and Arithmetic Problem Set 9

5. Out of the 60 homework assignments that Hong was assigned, she completed eleven assignments for every one assignment she didn't complete. How many of 60 assignments did Hong not complete?

6. The ratio of biologists to chemists at a biochemistry convention is 4:7. Which of the following could not be the total number of scientists attending?

   F. 121

   G. 198

   H. 220

   J. 341

   K. 387

7. A total of 30 million votes were cast for two presidential candidates, Bushwick and Clintonian. If Clintonian won by a ratio of 7 to 5, then what was the number of votes cast for Bushwick?

   A. $2,500,000$

   B. $12,500,000$

   C. $15,000,000$

   D. $17,500,000$

   E. $18,000,000$

8. If m and n are integers, $5 < m < 20$ and $\dfrac{m}{n} = \dfrac{5}{7}$, how many possible values are there for n?

   F. One

   G. Two

   H. Three

   J. Four

   K. Five

DO YOUR FIGURING HERE

Ratios
Algebra and Arithmetic Problem Set 9

9. The candies in a certain basket are either red or blue. If the ratio of the number of red candies to the number of blue candies is $\frac{3}{5}$, each of the following could be the number of candies in the basket EXCEPT

   A. 16

   B. 18

   C. 24

   D. 32

   E. 40

10. If the degree measures of the angles of a triangle are in the ratio 3 : 5 : 7, what is the measure of the largest angle in degrees?

    F. 36

    G. 40

    H. 56

    J. 60

    K. 84

11. Fruit punch is made from water, sugar, and juice concentrate mixed in a ratio of 5 to 1 to 2, respectively. What fraction of the punch is juice concentrate?

    A. $\frac{1}{8}$

    B. $\frac{1}{5}$

    C. $\frac{1}{4}$

    D. $\frac{2}{5}$

    E. $\frac{3}{4}$

Ratios
Algebra and Arithmetic Problem Set 9

2. After the first term in a sequence of positive integers, the ratio of each term to the term immediately preceding it is 4 to 1. What is the ratio of the 7th term to the 4th term?

   F. 2 to 1

   G. 4 to 1

   H. 8 to 1

   J. 16 to 1

   K. 64 to 1

3. If the ratio of s to t is 6 to 7, which of the following could be the values of s and t?

   A. $s = 2, t = 5$

   B. $s = 3, t = \dfrac{7}{2}$

   C. $s = 3, t = 8$

   D. $s = 12, t = 16$

   E. $s = 36, t = 49$

4. Cookies are to be removed from a jar that contains 20 butter cookies and 20 chocolate cookies. What is the smallest number of butter cookies that could be removed so that the ratio of the butter cookies to the chocolate cookies left in the jar is 3 to 5?

5. For positive integers a, b, and c, if a is one half the value of b, and b is three times the value of c, what is the ratio of a^2 to c^2?

   A. 2 to 3

   B. 3 to 2

   C. 4 to 9

   D. 9 to 4

   E. 11 to 13

Ratios
Algebra and Arithmetic Problem Set 9

Answer Key

#	Answer	Frequency	Difficulty
1	D	popular	1
2	K	popular	2
3	B	popular	2
4	G	popular	2
5	5	popular	2
6	K	popular	2
7	B	popular	2
8	G	popular	3
9	B	popular	2
10	K	popular	2
11	C	popular	2
12	K	popular	3
13	B	popular	2
14	8	popular	3
15	D	popular	4

Rate, Time, and Distance
Algebra and Arithmetic Problem Set 10

1. A tugboat crosses a lake at 15 mph and returns along the same path at 10 mph. If there was no current and the entire trip took one hour, what is the distance across the lake, in miles?

   A. 2

   B. 3

   C. 5

   D. 6

   E. 12

2. What was the average speed of the boat from problem 1 to the nearest mile per hour?

   F. 4

   G. 6

   H. 12

   J. 13

   K. 14

3. A plane leaves New York City bound for Los Angeles, CA at a rate of 500 mph and returns along the same path at 600 mph in a total trip of 11 hours. What is the distance, in miles, between the two cities?

   A. 5

   B. 6

   C. 300

   D. 3000

   E. 6000

4. What was the average speed of the plane in the problem 3?

   F. 450

   G. 545

   H. 575

   J. 600

   K. 1272

DO YOUR FIGURING HERE

Rate, Time, and Distance
Algebra and Arithmetic Problem Set 10

5. How many minutes did the plane from problem 3 take to return?

    A. 200

    B. 300

    C. 360

    D. 600

    E. 720

DO YOUR FIGURING HERE

6. Kelly rides her bike to the store at an average speed of 10mph. If she shops for 2 hours then rides her bike home at the same pace, and the total trip lasts 4 hours. What is the distance, in miles, between Kelly's home and the store?

    F. 5

    G. 10

    H. 15

    J. 20

    K. 25

7. If Evan drives 120 miles at 80 mph and returns along the same path at 100 mph, which of the following is closest to his average speed?

    A. 80

    B. 84

    C. 89

    D. 90

    E. 150

8. A car drives 80 mph for the final 3 hours of a 16 hour road trip. If the car's average speed for the 16 hour trip was 60 mph, what was the average speed for the first 13 hours of the trip?

    F. 55.38

    G. 57.24

    H. 58.64

    J. 60

    K. 73.85

Rate, Time, and Distance
Algebra and Arithmetic Problem Set 10

. Jose is walking 24 miles. If he travels at 4 mph for the first half of the walk and 3 mph for the final half of the walk, how many hours would Jose save by riding his bike at an average speed of 8 mph?

   A. 3

   B. 4

   C. 5

   D. 6

   E. 7

0. Felix jogs down the beach at 6 mph for 100 minutes. If it takes Felix 3 hours to walk back, what is the rate, in mph, at which he walks back?

   F. 2

   G. $2\frac{2}{3}$

   H. 3

   J. $3\frac{1}{3}$

   K. $4\frac{1}{4}$

. Andrew drives to work at 55mph. The round trip takes 45 minutes and it takes him 6 minutes longer to get home from work than it takes him to get to work. Approximately what speed (in mph) does Andrew drive home at?

   A. 22.5

   B. 25.5

   C. 42

   D. 45

   E. 50

Rate, Time, and Distance
Algebra and Arithmetic Problem Set 10

12. Roxanne rides her bike uphill to the corner store at 6 mph. The round trip takes 20 minutes and due to the hill, it takes her 4 minutes longer to get to the store than it does to return home. How much faster is Roxanne traveling on the ride home?

F. 3 mph

G. 4 mph

H. 6 mph

J. 9 mph

K. 10 mph

13. Steve's lawnmower can mow 5 ft^2 of lawn per second when operating at low power, 7 ft^2 of lawn per second at half power and 10 ft^2 of lawn per second at full power. He ran it for 2 seconds on low, 5 seconds on medium and 15 seconds on high. What was his average mowing rate in square feet per second to the nearest hundredth?

A. 7.00 ft^2/sec

B. 7.33 ft^2/sec

C. 8.86 ft^2/sec

D. 9.00 ft^2/sec

E. 9.50 ft^2/sec

14. Dahlia drives 120 miles at a rate of 15 miles per hour. If she then returns the same distance at a rate of 25mph, what is her average speed for the entire trip in miles per hour?

F. $18\frac{3}{4}$

G. $19\frac{1}{2}$

H. $20\frac{1}{3}$

J. $21\frac{3}{4}$

K. $22\frac{1}{2}$

DO YOUR FIGURING HERE

Rate, Time, and Distance
Algebra and Arithmetic Problem Set 10

5. A running track is divided into 100-meter sections with two straightaways and two semicircles. Bob and Jane are in the middle of the same straightaway and begin running in the opposite direction. If Bob runs at 86 seconds per lap and Jane at 72 seconds per lap, how far will Jane have run when she first passes Bob?

    A. 39 meters

    B. 182 meters

    C. 200 meters

    D. 217 meters

    E. 250 meters

6. A car traveled 40 miles at an average speed of 80 miles per hour and then traveled the next 40 miles at an average speed of 20 miles per hour. What was the average speed, in miles per hour of the car for the 80 miles?

7. Kim drove from her home to work at an average speed of 60 miles per hour and returned home along the same route at an average speed of 40 miles per hour. If her total driving time for the trip was 1 hour, how many minutes did it take Kim to drive from her home to work?

    A. 20

    B. 24

    C. 30

    D. 36

    E. 40

DO YOUR FIGURING HERE

Answer Key

#	Answer	Frequency	Difficulty
1	D	rare	3
2	H	rare	2
3	D	rare	3
4	G	rare	2
5	B	rare	2
6	G	rare	2
7	C	rare	3
8	F	rare	4
9	B	rare	4
10	J	rare	3
11	C	rare	4
12	F	rare	4
13	C	rare	4
14	F	rare	3
15	D	rare	5
16	32	rare	3
17	B	rare	4

Charts and Graphs
Algebra and Arithmetic Problem Set 11

. According to the table below, which of the five schools has the median number of text books?

School	Total Number of Textbooks	Textbooks Per 100 Students
A	2000	100
B	18,000	150
C	3000	200
D	2500	250
E	6000	300

A. School A

B. School B

C. School C

D. School D

E. School E

DO YOUR FIGURING HERE

. The circle graph below shows the results of a survey in which 5000 people reported the type of music they like. How many more of these 5000 chose rock than chose jazz?

Favorite Music Genres

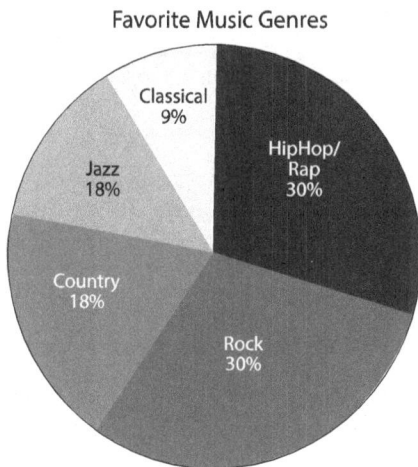

Classical 9%

HipHop/ Rap 30%

Jazz 18%

Country 18%

Rock 30%

F. 450

G. 600

H. 850

J. 1050

K. 1500

Charts and Graphs
Algebra and Arithmetic Problem Set 11

3. The partially completed chart below refers to 50 students' class choices for next year. 13 decided to take both a math and science class, 15 decided to drop both. How many of the 50 students decided to take a science class, but drop a math class?

		Math		Totals
		Yes	No	
Science	Yes	13		
	No		15	
Totals		30		50

A. 5

B. 17

C. 18

D. 20

E. 32

4. For Jim's birthday, he wanted to take himself and 49 friends and family members to a basketball game. He purchased 50 tickets, according to the discount rates shown below. On the day of the game, one of Jim's friends didn't show up, leaving him with one extra ticket. If each nondiscounted ticket cost $20, how much money did Jim save by purchasing the 50 tickets and using only 49 of them, rather than knowing his friend wouldn't show up and purchasing 49 tickets in advance?

Number of Tickets Purchased	Discount Per Ticket
1 to 9	$0
10 to 19	$2
20 to 29	$4
30 to 39	$6
40 to 49	$8
50 and up	$10

F. $2

G. $8

H. $20

J. $88

K. $108

The pie chart below shows the three sources of Laura's income. If 1/5 of the total amount of money is from child support and 1/3 is from waitressing, what is the percent of Laura's income earned from catering?

DO YOUR FIGURING HERE

Sources of Income for Laura

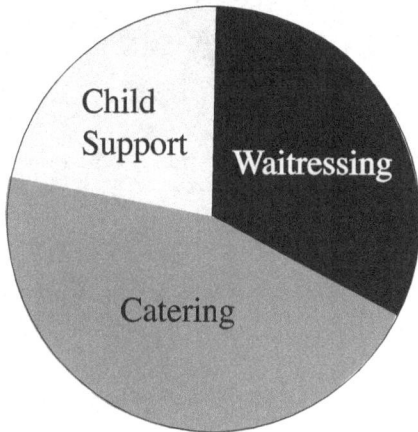

A. 25%

B. 33.3%

C. 46.6%

D. 53.3%

E. 66.6%

Charts and Graphs
Algebra and Arithmetic Problem Set 11

6. The chart below shows the results of a survey in which the average number of pets for 5000 people was calculated, where n represents the number of pets. How many people have more than 3 pets?

DO YOUR FIGURING HERE

Average Number of Pets

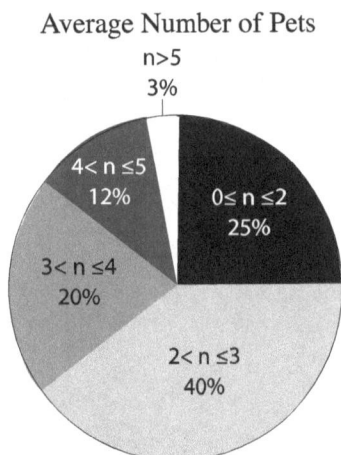

F. 150

G. 600

H. 750

J. 1000

K. 1750

7. The table below shows the results of subtracting the numbers $a, b, c,$ and d from each other. Each number in the body of the table is equal to the difference of the number at the top of the table minus the number to the left of the table. For example, $b - a = 6$. If $c = 12$, what is the value of $d - c + b - a$?

	a	b	c	d
a	0	6	9	13
b	-	0	3	7
c	-	-	0	4
d	-	-	-	0

A. -10

B. 10

C. 16

D. 22

E. 40

Charts and Graphs
Algebra and Arithmetic Problem Set 11

Use the table below to answer questions 8 through 10.

Suzanne would like to start a pie-baking company. Initially, she plans to make 500 of each of 3 types of pies. The table below shows the number of pounds of sugar, flour, and fruit she will need to fulfill this goal. Let s represent the price of 1 pound of sugar, f represent the price of 1 pound of flour, and r represent the price of 1 pound of fruit. All prices are in dollars.

Type of Pie	Pounds of Sugar	Pounds of Flour	Pounds of Fruit
Cherry	7	9	4.5
Pumpkin	8	8	6
Apple	4	7	9

How many pounds of flour are required to make 150 cherry pies?

F. 1.8

G. 2.4

H. 2.7

J. 3

K. 4.5

Refer to the table in question 8.

Suzanne has 25 pounds of sugar, 31 pounds of flour and 31.5 pounds of apples. What is the maximum number of apple pies Suzanne can make from the ingredients in stock?

A. 1,500

B. 1,750

C. 2,000

D. 2,200

E. 3,125

10. Refer to the table in question 8.
Which of the following expressions gives the price
of the sugar, flour and fruit required to make 500
cherry pies and 500 pumpkin pies?

F. $\dfrac{19}{s} + \dfrac{24}{f} + \dfrac{19.5}{r}$

G. $\dfrac{15}{s} + \dfrac{17}{f} + \dfrac{10.5}{r}$

H. $19s + 24f + 19.5r$

J. $15s + 17f + 10.5r$

K. $15s^2 + 17f^2 + 10.5r^2$

DO YOUR FIGURING HERE

Answer Key

#	Answer	Frequency	Difficulty
1	C	popular	2
2	G	popular	2
3	A	popular	3
4	J	popular	3
5	C	popular	2
6	K	popular	3
7	B	popular	3
8	H	popular	1
9	B	popular	1
10	J	popular	1

Percentage
Quick Drill

1. 9 is 18% of what number?

2. 18 is 30% of what number?

3. What is 850% of 16?

4. What percent of 60 is 420?

5. 420 is 14% of what number?

6. 48 is what percent of 64?

7. 1600 is what percent of 20?

8. 700 is what percent of 140?

9. What percent of 400 is 20% of 10,000?

10. 200% of 5 is what percent of 50?

11. What is the percent change from 800 to 600?

12. If there is an increase of 75% from 72, what is the final number?

Percentage
Quick Drill

13. If an increase of 150% gives a final number of 300, what is the initial number?

14. 304 is what number decreased by 24%?

15. What number increased by 14% is 980.4?

Percentage
Quick Drill

Answer Key

#	Answer
1	50
2	60
3	136
4	700%
5	3000
6	75%
7	8000%
8	500%
9	500%
10	20%
11	-25%
12	126
13	120
14	400
15	860

Percentage
Algebra and Arithmetic Problem Set 12

1. If 40 percent of f equals 200 percent of g, which of the following expresses g in terms of f?

   A. $g = 10\%$ of f

   B. $g = 20\%$ of f

   C. $g = 25\%$ of f

   D. $g = 40\%$ of f

   E. $g = 200\%$ of f

DO YOUR FIGURING HERE

2. If $x + 2y$ is equal to 150 percent of $6y$, what is the value of $\dfrac{x}{y}$?

3. Which of the following is equivalent to $\dfrac{1}{2}$ of 60 percent of 800?

   A. 30% of 400

   B. 30% of 1,200

   C. 60% of 400

   D. 60% of 800

   E. 80% of 600

4. In a class of 120 students, a vote for class president was split 60% to 40% between two candidates. How many votes would the "runner-up" need to take from the winner to win the majority?

   F. 12

   G. 13

   H. 24

   J. 25

   K. 48

5. If the positive number a is 250 percent of b, and if b is 20 percent of c, then a is what percent of c?

   A. 20%

   B. 25%

   C. 45%

   D. 50%

   E. 75%

112 LogicPrep ALGEBRA

2019-1

Percentage
Algebra and Arithmetic Problem Set 12

5. $\frac{1}{2}$ of 60 is what percent of 15?

    F. 3%

    G. 20%

    H. 30%

    J. 60%

    K. 200%

7. If 90 percent of c is equal to d percent of 30, where $d > 0$, what is the value of $\frac{c}{d}$?

    A. $\frac{1}{6}$

    B. $\frac{1}{3}$

    C. $\frac{1}{2}$

    D. 2

    E. 3

Which of the following is equal to $\frac{2}{m}$ percent of 20 for $m > 0$?

    F. $\frac{1}{m}$

    G. $\frac{2}{m}$

    H. $\frac{2}{5m}$

    J. $\frac{2m}{5}$

    K. $\frac{4m}{5}$

DO YOUR FIGURING HERE

Percentage
Algebra and Arithmetic Problem Set 12

9. The produce section of a supermarket currently has 350 apples. If the supermarket manager increases his produce by 40%, how many apples will he have in his produce section?

    A. 140

    B. 280

    C. 490

    D. 500

    E. 720

10. What percent of the integers from 1 to 20, inclusive, are neither primes nor multiples of 6?

    F. 40%

    G. 45%

    H. 50%

    J. 55%

    K. 60%

11. If the price of a house is $170,000 after depreciating 15% during a real estate slump, what was the original value of the house?

    A. $144,500

    B. $150,000

    C. $150,500

    D. $185,500

    E. $200,000

12. After receiving an 18% raise, Joe was making $70.80 a day at the local coffee store. How much was Joe making per day before his raise?

    F. $12.74

    G. $49.20

    H. $50

    J. $60

    K. $120

DO YOUR FIGURING HERE

Percentage
Algebra and Arithmetic Problem Set 12

3. The "Talk a Lot" telephone company plans on laying off 24% of its employees. If, after the layoff, there are 190 employees still working, how many employees did the "Talk a Lot" telephone company originally employ?

   A. 60

   B. 120

   C. 125

   D. 240

   E. 250

DO YOUR FIGURING HERE

4. If the population in Stillwater, NJ increased by 15% from 1998 to 1999, and the population was 16,100 in 1999, what was the population in 1998?

   F. 13,685

   G. 14,000

   H. 14,500

   J. 15,000

   K. 18,515

5. A family takes their 23-pound cat Fluffy to the vet and finds that she is 15% overweight. How many pounds overweight is Fluffy?

   A. 1

   B. 2

   C. 3

   D. 4

   E. 5

. A man went to the doctor and found that his cholesterol level was 60 mg too high. The doctor told him that he was 30% above the average cholesterol level for a man his age. How high is his current cholesterol level?

   F. 180 mg

   G. 200 mg

   H. 240 mg

   J. 260 mg

   K. 280 mg

Percentage
Algebra and Arithmetic Problem Set 12

17. The bear attacks in Alaska have gone down 8% since last year. If there were 4 fewer attacks this year, how many bear attacks were there in Alaska this year?

    A. 40

    B. 46

    C. 50

    D. 54

    E. 58

18. A candy bar company claims that all of their 6-oz. candy bars will be "50% bigger". How many ounces will the new candy bars weigh?

    F. 7

    G. 8

    H. 9

    J. 10

    K. 11

19. There are 25 more children than adults in attendance at a rock show. If there are a adults, then, in terms of a what percent of those in attendance at the rock show are children?

    A. $\dfrac{a}{a+25}\%$

    B. $\dfrac{a}{2a+25}\%$

    C. $\dfrac{a+25}{2a+25}\%$

    D. $\dfrac{100a}{2a+25}\%$

    E. $\dfrac{100\left(a+25\right)}{2a+25}\%$

DO YOUR FIGURING HERE

Percentage
Algebra and Arithmetic Problem Set 12

0. In an election, 1.2 million votes were cast and each vote was for either Jackson or Billings. Jackson received 12,000 more votes than Billings. What percent of the 1.2 million votes were cast for Jackson?

   F. 50.05%

   G. 50.1%

   H. 50.5%

   J. 51%

   K. 55%

1. A poll was taken of a population of teenagers, some who drive cars and some who do not. 52% of the teenage drivers said they prefer Mobil over Sunoco when asked "Which gas station do you prefer?" If 18% of the questioned teenagers are drivers and 864 of these drivers prefer Sunoco, how many teenagers took part in this poll?

Percentage
Algebra and Arithmetic Problem Set 12

Answer Key

#	Answer	Frequency	Difficulty
1	B	popular	3
2	7	popular	3
3	C	popular	2
4	G	popular	3
5	D	popular	3
6	K	popular	2
7	B	popular	3
8	H	popular	4
9	C	popular	1
10	G	popular	1
11	E	popular	1
12	J	popular	1
13	E	popular	1
14	G	popular	1
15	C	popular	2
16	J	popular	2
17	B	popular	2
18	H	popular	1
19	E	popular	2
20	H	popular	3
21	10,000	popular	4

Change in Percentage
Algebra and Arithmetic Problem Set 13

DO YOUR FIGURING HERE

1. The population at a college campus increased from 12,000 to 15,000, by what percent did the population increase?

   A. 20%

   B. 30%

   C. 22.5%

   D. 25%

   E. 35%

2. The pit bull population in Jersey City has increased from 1,200 to 4,800 in the past 5 years. By what percent did the population increase?

   F. 240%

   G. 280%

   H. 300%

   J. 320%

   K. 400%

3. After driving a car originally purchased for $60,000, the owner sold the car for $14,400. By what percent did the value of the car decrease?

   A. 24%

   B. 30%

   C. 36%

   D. 42%

   E. 76%

4. The average pay for a nurse in a certain hospital in 1995 was $38,500 a year. If the average pay for a nurse increased to $43,120 a year in 2000, by what percent did the average yearly pay for a nurse increase?

   F. 8%

   G. 10%

   H. 11%

   J. 12%

   K. 16%

Change in Percentage
Algebra and Arithmetic Problem Set 13

5. A guitar was purchased for $2,000 and later sold for $3,200. By what percent did the value of the guitar increase?

    A. 30%

    B. 40%

    C. 50%

    D. 60%

    E. 70%

DO YOUR FIGURING HERE

6. Minimum wage recently increased from $6.00 per hour to $6.60 per hour. By what approximate percent did the hourly wage increase?

    F. 6%

    G. 10%

    H. 12%

    J. 14%

    K. 16%

7. A book company's sales increased $x\%$ from $4,300 per month to $5,590 per month. What is the value of x?

    A. 23

    B. 30

    C. 40

    D. 45

    E. 50

8. If the price of gold per ounce dropped from $90 to $72, by what percent did the value of gold decrease?

    F. 18%

    G. 20%

    H. 24%

    J. 30%

    K. 36%

Change in Percentage
Algebra and Arithmetic Problem Set 13

The price of Apple stock dropped $x\%$ from $9.60 a share to $7.20 a share. What is the value of x?

DO YOUR FIGURING HERE

0. A dress company is having a "buy one get one half-off" sale this weekend. What percent do you save by purchasing two dresses during the sale?

   F. 150%

   G. 20%

   H. 25%

   J. 50%

   K. 75%

1. The value of a baseball card increases $y\%$ from $160 to $184. What is the value of y?

2. Boogie Down Night Club was originally purchased for $500,000 before being renovated. After $300,000 in renovations, Boogie Down Night Club was sold for $1,200,000. Approximately what percent profit did the investors make by purchasing, renovating, and selling Boogie Down Night Club?

   F. 20%

   G. 30%

   H. 50%

   J. 80%

   K. 150%

3. If the cost of printing a certain book increased 18% to $10.62 per book, what was the original price of printing per book?

   A. $1.91

   B. $8.70

   C. $8.80

   D. $9.00

   E. $12.53

Change in Percentage
Algebra and Arithmetic Problem Set 13

14. A hurricane displaced 43% of the dolphins along the coast of Florida, leaving 1425 dolphins. How many dolphins were there along the coast of Florida before the hurricane?

    F. 812

    G. 2500

    H. 3314

    J. 3400

    K. 3500

DO YOUR FIGURING HERE

15. The price of a painting was first increased by 20% and then the new price was decreased by 25%. The final price was what percent of the initial price?

    A. 80%

    B. 90%

    C. 95%

    D. 105%

    E. 110%

16. The price of a car is d dollars. A sales tax of 8 percent is added to the price of the car to obtain the final cost. If d is an integer, then the final cost, in dollars, is d times what number?

17. The cost of gas reached $3.78 per gallon last week, which was an increase of 5% from the previous week. What was the cost per gallon of gasoline the previous week?

Change in Percentage
Algebra and Arithmetic Problem Set 13

18. The number of students in the junior class at Newton High School first increased by 20% and then decreased by 5% during the year leaving 456 students in the class. How many students were in the class at the beginning of the year?

   F. 361

   G. 397

   H. 400

   J. 420

   K. 576

19. The employee discount at A and F clothing was 20% off of the "sale price" of all clothing during a "Wacky Weekend Sale". If an employee pays $36 for a sweater that was marked down 25%, what was the original price of the sweater?

   A. $52.20

   B. $54.00

   C. $60.00

   D. $76.40

   E. $81.00

DO YOUR FIGURING HERE

Change in Percentage
Algebra and Arithmetic Problem Set 13

Answer Key

#	Answer	Frequency	Difficulty
1	D	popular	1
2	H	popular	3
3	E	popular	3
4	J	popular	3
5	D	popular	3
6	G	popular	3
7	B	popular	3
8	G	popular	3
9	25	popular	3
10	H	popular	3
11	15	popular	3
12	H	popular	4
13	D	popular	2
14	G	popular	3
15	B	popular	3
16	1.08	popular	4
17	$3.60	popular	3
18	H	popular	4
19	C	popular	5

Mean, Median, and Mode
Quick Drill

Find the mean, median, and mode for the following
set:
3, 8, 2, 6, 8, 9

4. Find the mean, median, and mode for the following
    set:
    16, 32, 20, 22, 31, 40, 15

Find the mean, median, and mode for the following
set:
-31, 0, 42, -2, 8, -15, 22

5. Find the mean, median, and mode for the following
    set:
    $6\sqrt{3}, -\sqrt{2}, 4, 13, -\sqrt{3}, 4$

Find the mean, median, and mode for the following
set:
89, 90, 91, 87, 99, 88, 72, 90, 87

6. Find the mean, median, and mode for the following
    set:
    16.02, 38.41, 0.32, 15.71, 5.81, 0.42

Mean, Median, and Mode
Quick Drill

Answer Key

#	Answer
1	Mean: 6 Median: 7 Mode: 8
2	Mean: 3.4 Median: 0 Mode: None
3	Mean: 88.1 Median: 89 Mode: 87 and 90
4	Mean: 25.1 Median: 22 Mode: None
5	Mean: 4.4 Median: 4 Mode: 4
6	Mean: 12.78 Median: 10.76 Mode: None

Mean, Median, and Mode
Algebra and Arithmetic Problem Set 14

The average (arithmetic mean) of 7, 17, and a is 17. What is the value of a?

A. 17

B. 24

C. 27

D. 30

E. 51

DO YOUR FIGURING HERE

Sam, George, and Alan own a total of 177 CDs. If George owns 85 of them, what is the average (arithmetic mean) of the number of CDs owned by Sam and Alan?

F. 30

G. 46

H. 50

J. 88.5

K. 92

$$16 - 2a, 16, 16 + 2a$$

What is the average (arithmetic mean) of the 3 quantities in the list above?

A. 4

B. 16

C. 24

D. $4 + \dfrac{2a}{3}$

E. $4 - \dfrac{2a}{3}$

For which of the following lists of 7 numbers is the average (arithmetic mean) greater than the median?

F. 4, 5, 6, 9, 13, 14, 15

G. 5, 6, 7, 9, 10, 11, 12

H. 4, 5, 6, 9, 10, 11, 12

J. 6, 7, 8, 9, 10, 11, 12

K. 6, 7, 8, 9, 10, 10, 10

Mean, Median, and Mode
Algebra and Arithmetic Problem Set 14

5. If the average of 3 and a is 9 and the average of 5 and b is 6, what is the average of a and b?

  A. 6

  B. 7

  C. 9

  D. 11

  E. 15

DO YOUR FIGURING HERE

6. Fernando has an average of 90 on his first 4 chemistry quizzes. What does he have to score on his fifth quiz to bring his average up to a 92?

  F. He can't raise his average to a 92.

  G. 94

  H. 95

  J. 96

  K. 100

7. What is the average of eight hundredths and five tenths?

  A. .029

  B. .29

  C. 2.9

  D. 29

  E. 40

8. If the average (arithmetic mean) of 4, p, and q is 6, what is the value of $p + q$?

  F. 2

  G. 3

  H. 6

  J. 7

  K. 14

Mean, Median, and Mode
Algebra and Arithmetic Problem Set 14

DO YOUR FIGURING HERE

. Jay Leno has 4 cars with a top speed of 190 mph, 3 cars with a top speed of 200 mph and 10 cars with a top speed of 204 mph. What is the average (arithmetic mean) top speed of all 17 of his cars?

   A. 198 mph

   B. 200 mph

   C. 201 mph

   D. 202 mph

   E. 203 mph

0.
$$a, 2a, 3a, 4a, b$$
   If the average (arithmetic mean) of the 5 terms above is $3a$, what is b in terms of a?

   F. a

   G. $\dfrac{a}{3}$

   H. $3a$

   J. $5a$

   K. $15a$

1. If the average of 2 numbers is 35 and the smaller number is three fourths of the larger number, what is the smaller number?

   A. 20

   B. 30

   C. 40

   D. 50

   E. 60

2. What is the sum of 7 consecutive integers whose mean is 18?

   F. 77

   G. 92

   H. 111

   J. 118

   K. 126

Mean, Median, and Mode
Algebra and Arithmetic Problem Set 14

13. What is the average of $8, a, 15$ and b if $a + b = 22$?

   A. $\dfrac{45}{4}$

   B. $\dfrac{46}{4}$

   C. $\dfrac{48}{4}$

   D. $\dfrac{52}{4}$

   E. $\dfrac{45}{3}$

DO YOUR FIGURING HERE

14.
$$5, 8, 9, 10, 13, 14$$
The number c is to be added to the list above. If c is an integer, which of the following could be the median of the new list of seven numbers?

       I. 9
      II. 9.5
      III. 10

   F. I only

   G. I and II only

   H. III only

   J. I and III only

   K. I, II, and III

15. If $2 \le x \le 3$ and $1 \le y \le 6$, what is the lowest possible average of $\dfrac{1}{x}$ and y^2?

   A. $\dfrac{1}{3}$

   B. $\dfrac{2}{3}$

   C. 1

   D. $\dfrac{3}{2}$

   E. 2

Mean, Median, and Mode
Algebra and Arithmetic Problem Set 14

6. The average (arithmetic mean) of j, k, l, m, and n is equal to the median of j, k, and m. If $0 < j < k < m$, which of the following must be equal to k?

F. $\dfrac{(j+m)}{2}$

G. $\dfrac{(j+m)}{3}$

H. $\dfrac{(j+l+m+n)}{5}$

J. $\dfrac{(j+l+m+n)}{4}$

K. $\dfrac{(j-l-m-n)}{4}$

7. The average (arithmetic mean) of 4 numbers is a. If one of the numbers is b and another one of the numbers is c, what is the average of the 2 remaining numbers in terms of a, b, and c?

A. $\dfrac{a}{4}$

B. $\dfrac{(2b+2c-a)}{4}$

C. $\dfrac{(2a-b+c)}{4}$

D. $\dfrac{(4b+4c-a)}{2}$

E. $\dfrac{(4a-b-c)}{2}$

8. A car traveled 15 miles at an average speed of 30 miles per hour and then traveled the next 15 miles at an average speed of 60 miles per hour. What was the average speed, in miles per hour, of the car for the 30 miles?

F. 30

G. 40

H. 45

J. 50

K. 60

Mean, Median, and Mode
Algebra and Arithmetic Problem Set 14

19. Green Berry purchased a number of soft-serve yogurt machines: 7 costing $2,500 each, 10 costing $1,800 each and t costing $1,500 each, where t is a positive even integer. If the median price for all the yogurt machines purchased by Green Berry was $1,800, what is the greatest possible value of t?

DO YOUR FIGURING HERE

20. The average (arithmetic mean) of 5 numbers is a. If one of the numbers is b, what is the average of the remaining 4 numbers in terms of a and b?

    F. $\dfrac{9}{4}$

    G. $\dfrac{2a - b}{4}$

    H. $\dfrac{2a + b}{4}$

    J. $\dfrac{5a + b}{4}$

    K. $\dfrac{5a - b}{4}$

21. $5, 7, 8, 9, 7, 6, 6, x, 9, 6, 6$

    For the numbers listed above, the only mode is 6 and the median is 7. Each of the following could be the value of x EXCEPT:

    A. 6

    B. 8

    C. 9

    D. 10

    E. 11

Mean, Median, and Mode
Algebra and Arithmetic Problem Set 14

2. Beth was the only student in a class of 20 students that missed a test. If the class, without Beth, averaged a 93 on the test, what would Beth have to score to lower the class average to a 92?

   F. 53

   G. 60

   H. 73

   J. 87

   K. 90

3. A taxi costs $3 per minute for the first fifteen minutes and $.50 per minute for each additional minute. What is the average price per minute of a 20 minute taxi ride, rounded to the nearest cent?

   A. $1.75

   B. $2.00

   C. $2.25

   D. $2.38

   E. $3.00

4. The average age of a certain group of 30 women is 30 years. If 10 additional men are included in the group, then the average age of the 40 people is 28 years. What is the average age of the 10 men?

   F. 20

   G. 22

   H. 24

   J. 26

   K. 28

5. The 20 students in Ms. Martin's first period class scored an average of 86 on a test. Her second period class of 30 students averaged an 81 on the same test. What is the average of all of Ms. Martin's students' test scores?

   A. 82

   B. 83

   C. 84

   D. 85

   E. 86

26. The various averages (arithmetic means) of three of the four numbers $a, b, c,$ and d are calculated, and these averages are arranged from least to greatest as follows.

The average of $a, d,$ and b
The average of $d, c,$ and a
The average of $b, d,$ and c
The average of $a, b,$ and c

What is the order of $a, b, c,$ and d, from greatest to least?

F. $b > c > a > d$

G. $b > d > c > a$

H. $a > b > d > c$

J. $c > b > a > d$

K. $c > a > d > b$

27. Rich makes \$30/hour at the factory from noon to 6 pm. He then makes \$24/hour working at the store from 6 pm to 8 pm. Finally, he spends \$19/hour on night school from 8 pm to 9:30 pm. In one day, what is Rich's average net income per hour between the hours of noon and 9:30?

A. \$18

B. \$21

C. \$22

D. \$24

E. \$26

28. A car has two different drive settings: fuel efficient and sport. When driving in sport mode, the car averages 16 miles per gallon, and when driving in fuel-efficient mode, the car averages 28 miles per gallon. If during a 12-hour trip, the car drives 4 hours in sport mode and 8 hours in fuel efficient mode, then what was the average miles per gallon of the trip?

F. 21

G. 22

H. 23

J. 24

K. 25

DO YOUR FIGURING HERE

Mean, Median, and Mode
Algebra and Arithmetic Problem Set 14

29. A hot dog vendor calculated that his average daily profit over a seven-day week was $110. While making his calculations, however, he incorrectly entered $110 instead of $120 for Monday and $105 instead of $130 for Tuesday. What was his actual average daily profit over the seven-day time period?

A. $105

B. $107

C. $115

D. $120

E. $130

30. The average grade in a class of t students is 84 and the average grade in a s students is 90. If the average of both classes is 88, what is the value of $\dfrac{t}{s}$?

DO YOUR FIGURING HERE

Mean, Median, and Mode
Algebra and Arithmetic Problem Set 14

Answer Key

#	Answer	Frequency	Difficulty
1	C	popular	1
2	G	popular	2
3	B	popular	2
4	F	popular	3
5	D	popular	2
6	K	average	2
7	B	popular	1
8	K	popular	1
9	B	average	1
10	J	popular	2
11	B	popular	3
12	K	popular	2
13	A	popular	2
14	J	popular	4
15	B	popular	4
16	J	popular	3
17	E	popular	4
18	G	popular	4
19	16	popular	4
20	K	popular	3
21	A	popular	4
22	H	average	2
23	D	average	2
24	G	average	1
25	B	average	1
26	J	popular	5
27	B	average	3
28	J	average	3
29	C	average	3
30	$\frac{1}{2}$	average	4

Combinatorics
Algebra and Arithmetic Problem Set 15

Joe has four different pairs of shoes and seven differently colored laces. How many different shoe-lace combinations can he make?

A. 4!

B. 7!

C. 11

D. 28

E. 35

DO YOUR FIGURING HERE

Sweater-vests come in 5 sizes: small, medium, large, X-large and XX-large and 4 different styles: plaid, argyle, striped and solid color. How many types of sweater-vests can be ordered?

F. 9

G. 9!

H. 20

J. 20!

K. 29

How many positive four-digit integers can be made from the numbers 2, 3, 4, and 6, where each digit must be used once and only once?

A. 1

B. 4

C. 24

D. 120

E. 240

The Oak Park Little League has 5 teams. For the first round of the playoffs, each team plays each other team once. How many games are played in this round?

F. 5

G. 10

H. 15

J. 120

K. 240

Combinatorics
Algebra and Arithmetic Problem Set 15

5. How many positive three-digit integers have the tens place equal 2 and the ones digit equal 1?

   A. 1

   B. 2

   C. 3

   D. 9

   E. 10

6. Jeff mixes and matches all of his pants, t-shirts, and sneakers to make different outfits. If he can make a total of 60 different outfits, each consisting of one pair of pants, one t-shirt, and one pair of sneakers, which of the following could NOT be the number of t-shirts that Jeff has?

   F. 2

   G. 4

   H. 6

   J. 8

   K. 10

7. Latisha has red, yellow, and blue scarves. She also has red, yellow and blue sunglasses and red, yellow and blue pants. If she wants to wear all three colors on any given day, then how many distinct outfits can she wear?

   A. 1

   B. 3

   C. 6

   D. 9

   E. Infinitely many

8. How many unique ordered pairs (x, y) can be made such that x is an even integer greater than 5 and less than 11 and Y is an integer greater than 5 and less than 7?

   F. 3

   G. 5

   H. 11

   J. 15

   K. 45

DO YOUR FIGURING HERE

Combinatorics
Algebra and Arithmetic Problem Set 15

How many 2 digit positive integers have only odd integers as digits?

A. 0

B. 2

C. 25

D. 49

E. 50

DO YOUR FIGURING HERE

. Sean has a black hat, a white hat, and a blue hat. He also has three scarves, one black, one white, and one blue and three sweaters, one black, one white and one blue. Sean wants to wear all three colors each day. How many different possibilities does he have?

F. 6

G. 9

H. 12

J. 18

K. 27

. Lizzy mixes her spices, meats and vegetables to make 28 different soups, each with one spice, one meat and one vegetable. Which cannot be the number of spices she has?

A. 1

B. 2

C. 3

D. 4

E. 28

. A snack machine has buttons A, B, C, D, E, F and numbers $1, 2, 3, 4, 5,$ and 6. A selection is made with a letter and then a two-digit number. How many selections can be made?

F. 216

G. 396

H. 594

J. 720

K. It cannot be determined with the data given.

Combinatorics
Algebra and Arithmetic Problem Set 15

13. How many positive four digit-integers can be made where the first digit must be 9, the last digit cannot be 9 and digits can be repeated?

    A. 90

    B. 900

    C. 999

    D. 1000

    E. 9999

14. There are 4 roads from Newton to Sparta and 5 roads from Sparta to Stillwater. If Helen drives from Newton to Stillwater and back, passes through Sparta in both directions, and does not travel any road twice, how many different routes for the trip are possible?

    F. 80

    G. 120

    H. 200

    J. 240

    K. 360

15. In a baseball league with 6 teams, each team plays exactly 2 games with each of the other 5 teams in the league. What is the total number of games played in the league?

    A. 6

    B. 18

    C. 30

    D. 32

    E. 36

16. Any 2 points determine a line. If there are 8 points in a plane, no 3 of which lie on the same line, how many lines are determined by pairs of these 8 points?

    F. 12

    G. 16

    H. 20

    J. 28

    K. 36

Combinatorics
Algebra and Arithmetic Problem Set 15

7. The Power House Electric company will send a team of 4 electricians to work on a certain job. The company has 5 experienced electricians and 5 trainees. If a team consists of 1 experienced electrician and 3 trainees, how many such teams are possible?

DO YOUR FIGURING HERE

8. A baseball league has 30 teams and each team must play each other team three times. What is the total number of baseball games that are played in this league?

F. 870

G. 900

H. 1305

J. 1740

K. 2610

Combinatorics
Algebra and Arithmetic Problem Set 15

Answer Key

#	Answer	Frequency	Difficulty
1	D	average	1
2	H	average	1
3	C	average	1
4	G	average	2
5	D	average	1
6	J	average	1
7	C	average	2
8	F	average	2
9	C	average	2
10	F	average	1
11	C	average	1
12	F	average	2
13	B	average	3
14	J	average	3
15	C	average	3
16	J	average	3
17	50	average	3
18	H	average	4

Probability
Algebra and Arithmetic Problem Set 16

1. Joe accidentally left a package on one of 120 tables. If exactly 33 of those tables are blue, what is the probability that the package will be on a blue table?

 A. $\dfrac{1}{120}$

 B. $\dfrac{3}{12}$

 C. $\dfrac{11}{40}$

 D. $\dfrac{10}{12}$

 E. 1

2. If 1 out of 5 pizzas in the factory have unsafe amounts of lead in the cheese, what is the probability that a pizza selected at random will be safe to eat?

 F. $\dfrac{1}{5}$

 G. $\dfrac{2}{5}$

 H. $\dfrac{3}{5}$

 J. $\dfrac{4}{5}$

 K. 1

3. There are 19 coins in a bag: four nickels, five pennies, six dimes and the rest are quarters. If a coin is drawn at random, what is the probability that it is a quarter?

 A. $\dfrac{3}{19}$

 B. $\dfrac{4}{19}$

 C. $\dfrac{5}{19}$

 D. $\dfrac{5}{14}$

 E. $\dfrac{4}{14}$

4. Of the dresses in Mia's closet, 4 are pink. When she randomly picks a dress, the probability that it is pink is $\frac{2}{5}$. How many dresses does she have?

   F. 2

   G. 5

   H. 10

   J. 20

   K. 40

DO YOUR FIGURING HERE

5. A ball is to be chosen randomly from a basket of balls. The probability that it will be a baseball is $\frac{1}{7}$. Which cannot be a number of balls in the basket?

   A. 7

   B. 21

   C. 49

   D. 100

   E. 140

6. A bag contains only buttons, marshmallows and pennies. The probability of choosing a marshmallow is $\frac{1}{5}$ and the probability of choosing a penny is $\frac{1}{7}$. Which could be the number of items in the bag?

   F. 5

   G. 7

   H. 35

   J. 75

   K. 100

Probability
Algebra and Arithmetic Problem Set 16

Of 24 ice cream scoops, the most common flavor is chocolate. What is the probability that a scoop of ice cream chosen randomly is not chocolate?

A. 0

B. $\dfrac{11}{24}$

C. $\dfrac{12}{24}$

D. $\dfrac{13}{24}$

E. Cannot be determined

DO YOUR FIGURING HERE

In a bowl there are 21 pieces of paper, each with a different integer from 1 to 21 (inclusive) on it. If one is drawn at random, what is the probability that it will be prime?

F. $\dfrac{4}{21}$

G. $\dfrac{5}{21}$

H. $\dfrac{6}{21}$

J. $\dfrac{1}{3}$

K. $\dfrac{8}{21}$

$$1\ 2\ x\ 6\ 7\ 9$$

When a number is chosen randomly from the above numbers, the probability that it will be less than 5 is $\dfrac{1}{3}$. Which can be a value for x?

A. 1

B. 2

C. 3

D. 4

E. 5

10. A can has 5 pieces of salmon, 4 shrimp, 3 pieces of bass, 2 pieces of marlin and 1 piece of tuna. The probability of Joe picking his favorite fish is $\frac{1}{3}$. What is Joe's favorite fish?

F. Salmon

G. Shrimp

H. Bass

J. Marlin

K. Tuna

11. If a number is chosen randomly from the set $\{-10, -9, 8, 9, 10\}$, what is the probability that it is a member of the solution set of both $2x > 0$ and $\frac{x}{2} < 0$?

A. 0

B. $\frac{1}{5}$

C. 1

D. $\frac{1}{2}$

E. No solution

12. $XOOX$

The figure above is four bikes that will be assigned to four bikers. Joe and Brian are two of the bikers. What is the probability that both will receive an X bike?

F. $\frac{1}{16}$

G. $\frac{1}{12}$

H. $\frac{1}{9}$

J. $\frac{1}{6}$

K. $\frac{1}{3}$

DO YOUR FIGURING HERE

Probability
Algebra and Arithmetic Problem Set 16

3. In a group of 10 boys and 12 girls, 5 boys and 1 girl speak French. If a boy and a girl are chosen at random, what is the probability that at least one of them speaks French?

A. $\dfrac{1}{24}$

B. $\dfrac{1}{12}$

C. $\dfrac{1}{10}$

D. $\dfrac{5}{12}$

E. $\dfrac{13}{24}$

4. A box contains brown plastic poker chips, red plastic poker chips and silver metal poker chips. There are 3 times as many metal poker chips as there are plastic ones. If a chip is randomly picked, the probability that it is red is 4 times the probability that it is brown. If there are 2 brown chips in the box, how many chips are there in total?

F. 2

G. 10

H. 38

J. 40

K. 48

5. $\qquad A, B, C, D, E, F, G, H$
A list has all possible three-letter arrangements of the letters above with the first letter being H and one of the other letters being F. If no letter is reused in each arrangement and one three-letter arrangement is randomly chosen, then what is the probability of choosing HFD?

A. $\dfrac{1}{16}$

B. $\dfrac{1}{12}$

C. $\dfrac{1}{8}$

D. $\dfrac{1}{7}$

E. $\dfrac{1}{6}$

DO YOUR FIGURING HERE

Probability
Algebra and Arithmetic Problem Set 16

16. A box contains 1 red, 2 white and 3 blue marbles. If 2 marbles are to be drawn randomly, what is the probability that at least one will be white?

    F. $\dfrac{1}{6!}$

    G. $\dfrac{2}{6!}$

    H. $\dfrac{1}{6}$

    J. $\dfrac{1}{3}$

    K. $\dfrac{3}{5}$

17. The integers 1 through 6 appear on the six faces of a cube, one on each face. If three such cubes are rolled, what is the probability that the sum of the numbers on the top faces is 17 or 18?

    A. $\dfrac{1}{54}$

    B. $\dfrac{1}{48}$

    C. $\dfrac{1}{32}$

    D. $\dfrac{1}{16}$

    E. $\dfrac{1}{12}$

18. Matthew has a large bucket of 900 ping pong balls, all of equal size and weight, each labeled a different number from 101 to 1000. If Matthew selects one ball at random from the bucket, what is the probability that the ball is labeled with a number that is a perfect square?

    F. $\dfrac{31}{1000}$

    G. $\dfrac{21}{1000}$

    H. $\dfrac{20}{1000}$

    J. $\dfrac{20}{900}$

    K. $\dfrac{21}{900}$

DO YOUR FIGURING HERE

Probability
Algebra and Arithmetic Problem Set 16

Answer Key

#	Answer	Frequency	Difficulty
1	C	average	1
2	J	average	1
3	B	average	2
4	H	average	1
5	D	average	1
6	H	average	1
7	E	average	2
8	K	average	2
9	E	average	2
10	F	average	2
11	A	average	3
12	J	average	3
13	E	average	4
14	J	average	4
15	B	average	4
16	K	average	4
17	A	average	4
18	K	average	4

Exponents I
Quick Drill

1. $\left(x^5\right)^6 =$

2. $\dfrac{x^4}{x^3} =$

3. $\dfrac{m^6}{m^2} =$

4. $x^2 x^4 =$

5. $\dfrac{r^3}{r^3} =$

6. $g\left(gh\right)^2 =$

7. $\left(9^2\right)^2 =$

8. $\left(3^x\right)^{4y} =$

9. $\left(3x^2\right)^3 =$

10. $(x-2)^2 =$

11. $x^5 x^6 =$

Exponents I
Quick Drill

12. $\left(x^3\right)\left(x^y\right) =$

13. $\dfrac{d^3}{2d^2} =$

14. $\left(x^8\right)^y =$

15. Solve for a:
$$\dfrac{a^2}{a^3} = 3$$

16. Solve for b:
$$\dfrac{b^5}{b^3} = 49$$

17. Solve for k:
$$2k^2 + 3k^2 + 10k = 0$$

18. Solve for x:
$$3^{5x} = 243$$

19. Solve for k:
$$\dfrac{k^2}{2} = \text{-}\dfrac{32}{k}$$

20. Solve for x:
$$\dfrac{2x^2y^2 + 3x^2 + 6x}{2} = (xy)^2$$

Exponents I
Quick Drill

Answer Key

#	Answer
1	x^{30}
2	x
3	m^4
4	x^6
5	1
6	$g^3 h^2$
7	6561
8	3^{4xy}
9	$27x^6$
10	$x^2 - 4x + 4$
11	x^{11}
12	x^{3+y}
13	$\dfrac{d}{2}$
14	x^{8y}
15	$a = \dfrac{1}{3}$
16	$b = \pm 7$
17	$k = \text{-}2, 0$
18	$x = 1$
19	$k = \text{-}4$
20	$x = \text{-}2, 0$

Exponents II
Quick Drill

1. $x^{-1} =$

2. $\left(\dfrac{x^9}{x^3}\right)^{\frac{1}{2}} =$

3. $\dfrac{x^{-2}y^4}{x^3y^{-1}} =$

4. $\dfrac{a^{\frac{3}{2}}}{a^{\frac{2}{3}}} =$

5. $x^{-2} =$

6. $\left(x^6\right)^{\frac{1}{2}} =$

7. $\left(x^6\right)^{\frac{2}{3}} =$

8. $\dfrac{v^{-2}k^9}{k^{-2}v^2} =$

9. Solve for z:
$3\left(2z\right)^2 = 108$

10. Express as a fraction with positive exponents
$\dfrac{a^{-3}b^{-4}c^2}{a^4b^5c}$

Answer Key

#	Answer
1	$\dfrac{1}{x}$
2	x^3
3	$\dfrac{y^5}{x^5}$
4	$a^{\frac{5}{6}}$
5	$\dfrac{1}{x^2}$
6	x^3
7	x^4
8	$\dfrac{k^{11}}{v^4}$
9	$z = \pm 3$
10	$\dfrac{c}{a^7 b^9}$

Scientific Notation
Quick Drill

1. What is 6.28×10^7 in standard notation?:

2. What is 820,000 in scientific notation?

3. What is 3.02×10^4 in standard notation?

4. What is -704,500,000,000 in scientific notation?

5. What is 0.0023 in scientific notation?

6. What is 1.06×10^{-8} in standard notation?

7. Give your answer in scientific notation.
$7.1 \times 10^4 + 1.04 \times 10^4 =$

8. Give your answer in scientific notation.
$8.3 \times 10^6 + 2.0 \times 10^5 =$

9. Give your answer in scientific notation.
$\left(6.4 \times 10^{-3}\right)\left(2.0 \times 10^{-5}\right)=$

10. Give your answer in scientific notation.
$3.38 \times 10^{-3} + 1.02 \times 10^{-3} =$

11. Give your answer in scientific notation.
$2.12 \times 10^4 + 1.3 \times 10^3 =$

Scientific Notation
Quick Drill

12. Give your answer in scientific notation.
$$(4.6 \times 10^8)(3.2 \times 10^2) =$$

13. Give your answer in scientific notation.
$$2.2 \times 10^3 - 1500$$

14. Give your answer in scientific notation.
$$6.1 \times 10^{-8} - 1.6 \times 10^{-9} =$$

15. Give your answer in scientific notation.
$$\frac{3.0 \times 10^4}{1.5 \times 10^2} =$$

16. Give your answer in scientific notation.
$$\frac{5.4 \times 10^3}{1.8 \times 10^{-2}} =$$

17. Give your answer in scientific notation.
$$7.2 \times 10^4 - 1.1 \times 10^3 =$$

18. Give your answer in scientific notation.
$$9.8 \times 10^{-3} + 0.0062 =$$

19. Give your answer in scientific notation.
$$3.2 \times 10^4 - 1.0 \times 10^2 =$$

20. Give your answer in scientific notation.
$$(3200)(2.0 \times 10^4) =$$

Scientific Notation
Quick Drill

Answer Key

#	Answer
1	62,800,000
2	8.2×10^5
3	30,200
4	-7.045×10^{11}
5	2.3×10^{-3}
6	0.0000000106
7	8.14×10^4
8	8.5×10^6
9	1.28×10^{-7}
10	4.4×10^{-3}
11	2.25×10^4
12	1.472×10^{11}
13	7.0×10^2
14	5.94×10^{-8}
15	2.0×10^2
16	3.0×10^5
17	7.09×10^4
18	1.6×10^{-2}
19	3.19×10^4
20	6.4×10^7

Exponents
Algebra and Arithmetic Problem Set 17

1. When 850,000 is written as 8.5×10^a, what is the value of a?

   A. 1

   B. 2

   C. 3

   D. 4

   E. 5

DO YOUR FIGURING HERE

2. If $(x - 4)^2 = 36$, then x could be

   F. -10

   G. -6

   H. 2

   J. 4

   K. 10

3. If $a^2 - 36 = 0$, which of the following could be a value of a?

   A. -6

   B. -3

   C. 0

   D. 12

   E. 24

4. $3^4 = ?$

   F. 7

   G. 12

   H. 27

   J. 81

   K. 243

5. $(a^4)(a^3) = ?$

   A. a

   B. $7a$

   C. a^2

   D. a^7

   E. a^{12}

Exponents
Algebra and Arithmetic Problem Set 17

$\dfrac{43^9}{43^9} = ?$

F. $\dfrac{1}{43^{\frac{9}{8}}}$

G. $\dfrac{1}{43}$

H. 1

J. 43

K. $43^{\frac{9}{8}}$

DO YOUR FIGURING HERE

$\left(3^2\right)^9 = ?$

A. 54

B. 3^{11}

C. 3^{18}

D. 3^{29}

E. 3^{512}

$2^{-3} = ?$

F. -6

G. -1

H. $-\dfrac{3}{2}$

J. $-\dfrac{1}{8}$

K. $\dfrac{1}{8}$

$9^{\frac{3}{2}} = ?$

A. $\dfrac{1}{81}$

B. $\dfrac{1}{27}$

C. 10.5

D. 13.5

E. 27

10. If $x^4 = y^{14}z^8$, and if x, y, and z are positive numbers, then $x^2 =$

F. y^7z^8

G. $\dfrac{(y^7z^8)}{2}$

H. y^7z^4

J. $\dfrac{(y^7z^4)}{2}$

K. $\dfrac{(y^{14}z^8)}{2}$

DO YOUR FIGURING HERE

11. The amount of dirt that is shoveled out of a hole after t minutes is $4t + 2t^3$ pounds. How many pounds of dirt have been shoveled from the hole after 2 minutes?

A. 6

B. 10

C. 12

D. 16

E. 24

12. If $3^a = n$, where a is a positive integer, which of the following could be the value of n?

F. 1

G. 12

H. 36

J. 81

K. 100

13. If $12x^4 + 8x^2 + x = 9 + 8x^2 + 12x^4$, then $x =$

A. 1

B. 8

C. 9

D. 12

E. 24

Exponents
Algebra and Arithmetic Problem Set 17

4. $3^{\frac{-4}{2}} = ?$

   F. $\dfrac{-1}{2}$

   G. $\dfrac{-1}{6}$

   H. $\dfrac{1}{9}$

   J. 6

   K. 12

5. If $x^2 - x = 30$, which of the following is a possible value of $x^2 + x$?

   A. -30

   B. 30

   C. 42

   D. 60

   E. 870

6. If $3^a = 27$ and $a = \dfrac{b}{4}$, what is the value of b?

   F. 8

   G. 9

   H. 10

   J. 11

   K. 12

7. If x and y are integers such that $x^2 = 81$ and $y^4 = 81$, which of the following could be true?

           I. $x = -9$
          II. $x = 3$
          III. $x + y = 6$

   A. I only

   B. III only

   C. I and III only

   D. II and III only

   E. I, II, and III

18. If a and b are positive integers and $81^a = 3^b$, what is the value of $\dfrac{b}{a}$?

DO YOUR FIGURING HERE

    F. $\dfrac{1}{9}$

    G. $\dfrac{1}{4}$

    H. $\dfrac{1}{3}$

    J. 4

    K. 9

19. If x and y are positive integers and $8\left(2^x\right) = 2^y$, what is x in terms of y?

    A. $y - 3$

    B. $y - 1$

    C. y

    D. $y + 1$

    E. $y + 3$

20. If m is a positive integer, which of the following is NOT equal to $\left(3^3\right)^m$?

    F. $3^m\left(3^{2m}\right)$

    G. $9^m\left(3^m\right)$

    H. 3^{3m}

    J. 9^{2m}

    K. 27^m

21. Squaring the sum of b squared and 3 gives the same result as squaring the product of b and 3. Which of the following equations could be used to find all possible values of b?

    A. $b^4 - 9 = 3b^2$

    B. $b^4 + 9 = 9b^2$

    C. $\left(b^2 + 3\right)^2 = 3b^2$

    D. $\left(b^2 + 3\right)^2 = (3b)^2$

    E. $b^4 + 9 = 32^b$

Exponents
Algebra and Arithmetic Problem Set 17

2. If a and b are numbers such that $(a - 7)(b + 7) = 0$, what is the smallest possible value of $a^2 + b^2$?

   F. 0

   G. 7

   H. 14

   J. 49

   K. 98

3. Let the function f be defined by $f(x) = x^3 - 9$ and let the function g be defined by $g(x) = 2x^2$. If $f(a) = 55$, what does $g(a)$ equal?

   A. .5

   B. 4

   C. 16

   D. 32

   E. 64

4.
$$y < y^3 < y^2$$
For the inequality above, which MUST be true?

   F. $y > 0$

   G. $0 < y < 1$

   H. $-1 < y < 0$

   J. $y < 0$

   K. None of the above, the inequality is not possible

5. If $7^{k-3} + 7^2 = 98$, what is the value of k?

   A. 2

   B. 5

   C. 7

   D. 10

   E. 14

DO YOUR FIGURING HERE

Exponents
Algebra and Arithmetic Problem Set 17

26.
$$(x - 2j)^6 = (x - 2j)^2 a^4$$
In the equation above, x and a are positive numbers and $0 < 2j < x$. Which of the following must be equal to x?

F. a

G. a^4

H. $a^4 + 2j$

J. $a^4 - 2j$

K. $a + 2j$

27. Let $x \Omega y$ be defined by $x \Omega y = x - (x - y)^y$ for all positive integers x and y. What is the value of $(6 \Omega 1) \Omega 2$?

A. -1

B. 0

C. 2

D. 13

E. 134

28. If $x^2 + bx + 24 = (x - a)(x - 8)$ for all values of x and if b and a are constants, what is the value of b?

F. -11

G. -3

H. 3

J. 5

K. 11

29. If $a, b,$ and c are positive integers and $a^3 b^2 c^4 > a^3 b^3 c^3$, which of the following must be true?

I. $a < b$
II. $a < c$
III. $b < c$

A. I only

B. II only

C. III only

D. II and III only

E. I, II, III

DO YOUR FIGURING HERE

Exponents
Algebra and Arithmetic Problem Set 17

. Each of the following inequalities is true for some values of a EXCEPT:

F. $a^3 > a^2 > a$

G. $a^3 > a > a^2$

H. $a > a^2 > a^3$

J. $a^2 > a^3 > a$

K. $a^2 > a > a^3$

DO YOUR FIGURING HERE

. If $x^3 = y$, what does x^6 equal in terms of y?

A. $y^{\frac{9}{5}}$

B. y^2

C. y^4

D. y^5

E. y^6

. If x is a positive integer satisfying $x^6 = a$ and $x^{13} = b$, which of the following must be equal to x^{23}?

F. $b + 10$

G. $a^2 - 13$

H. $2b - \left(\dfrac{9}{2}\right)$

J. $\dfrac{a^6}{b}$

K. $a^2 - b$

33. If $k > 1$, which of the following must be less than k?

$$\text{I. } (k-1)^2$$
$$\text{II. } (k-2)^2$$
$$\text{III. } \frac{1}{(1-k)}$$

A. I only

B. II only

C. III only

D. I and II only

E. II and III only

DO YOUR FIGURING HERE

34. For all value of a, let a Δ be defined by $a\Delta = a^2 + 2$. Which of the following is equal to $(x\Delta)\,\Delta$?

F. $x^4 + 4$

G. $x^4 - x^2 + 2$

H. $x^4 + 4x + 4$

J. $x^4 + 6$

K. $x^4 + 4x^2 + 6$

35. If $k = x + 6$, what does $x^2 - 36$ equal, in terms of k?

A. $(k+6)^2$

B. $(k-6)^2$

C. $k(k+6)$

D. $k(k-6)$

E. $k(k-12)$

Exponents
Algebra and Arithmetic Problem Set 17

6.
$$H^4, H^3, \frac{1}{H}, \frac{1}{H^2}, \frac{1}{H^3}$$

If $-1 < H < 0$, what is the median of the five numbers in the list above?

F. $\dfrac{1}{H^3}$

G. $\dfrac{1}{H^2}$

H. $\dfrac{1}{H}$

J. H^3

K. H^4

7. If $a^{\frac{4}{3}} = b$, what does a^4 equal in terms of b?

A. $b^{\frac{1}{3}}$

B. b

C. b^3

D. b^4

E. b^6

8. If m and n are positive integers, which of the following is equivalent to $(2m)^{3n} - (2m)^n$?

F. $(2m)^{2n}$

G. $(2m)^n \left[(2m)^{4n} - 2m^n \right]$

H. $2m \left(m^{4n} - m^{2n} \right)$

J. $(2m)^n \left[(2m)^{2n} - 1 \right]$

K. $2m^{3n} \left(2m^{2n} - 1 \right)$

DO YOUR FIGURING HERE

39. A ball thrown upward from a height k feet with initial velocity j feet per second will reach a maximum height of $k + j^{\frac{1}{2}}$ feet. If the ball is thrown upward from a height of 5.5 feet with an intial velocity of 16 feet per second, what will its maximum height in feet?

   A. 5.5

   B. 5.75

   C. 9.5

   D. 12

   E. 15

DO YOUR FIGURING HERE

Exponents
Algebra and Arithmetic Problem Set 17

Answer Key

#	Answer	Frequency	Difficulty
1	E	popular	1
2	K	popular	3
3	A	popular	1
4	J	popular	1
5	D	popular	1
6	H	popular	1
7	C	popular	1
8	K	popular	2
9	E	popular	2
10	H	popular	2
11	E	popular	1
12	J	popular	2
13	C	popular	2
14	H	popular	2
15	C	popular	3
16	K	popular	2
17	C	popular	2
18	J	popular	3
19	A	popular	2
20	J	popular	2
21	D	popular	4
22	J	popular	2
23	D	popular	3
24	H	popular	3
25	B	popular	2
26	K	popular	3
27	B	popular	3
28	F	popular	3
29	C	popular	2
30	G	popular	3
31	B	popular	3
32	J	popular	4
33	C	popular	4
34	K	popular	3
35	E	popular	3
36	J	popular	4
37	C	popular	3
38	J	popular	4
39	C	popular	3

Radicals
Quick Drill

1. $\sqrt{81} =$

2. $\sqrt{96} =$

3. $\sqrt{120} =$

4. $\sqrt{200} =$

5. $\sqrt{250} =$

6. $\sqrt{80} =$

7. $\sqrt{60} =$

8. $\sqrt{256} =$

9. $\sqrt{27} =$

10. $\sqrt{180} =$

11. $\sqrt{240} =$

12. $\sqrt{x}\sqrt{x} =$

13. $\sqrt{x^3} =$

14. $\sqrt{k^2 k^{10}} =$

15. $\left(\dfrac{x^8}{x^2}\right)^{\frac{1}{2}} =$

16. $\sqrt{x^5} =$

17. $\left(x^8\right)^{\frac{1}{2}} =$

18. $\left(x^3\right)^{\frac{4}{5}} =$

19. $\sqrt{98x^2 y} =$

20. Solve for x:
$x\sqrt{x} = 8$

21. Solve for a:
$\dfrac{a}{\sqrt{a}} = 3$

22. $\sqrt[3]{x}\,\sqrt[3]{x} =$

Radicals
Quick Drill

23. $\dfrac{\sqrt{x^5}}{\sqrt[3]{x^2}} =$

24. $\dfrac{\sqrt[3]{8x^6y^3z}}{x^2y} =$

25. $\sqrt[3]{27a^3b^6} =$

26. $\sqrt[3]{x^2y^6z^4} =$

27. Express in simplest terms:
$$\dfrac{3\sqrt{2} + 4\sqrt{2}}{7\sqrt{3}}$$

28. $4^{\frac{1}{2}} + \left(\dfrac{8x^3}{27}\right)^{\frac{1}{3}} =$

29. Solve for x:
$$\left(2x^2\right)^{\frac{1}{5}} = \sqrt[5]{-2x + 4}$$

30. $\sqrt{432x} =$

Radicals
Quick Drill

Answer Key

#	Answer
1	9
2	$4\sqrt{6}$
3	$2\sqrt{30}$
4	$10\sqrt{2}$
5	$5\sqrt{10}$
6	$4\sqrt{5}$
7	$2\sqrt{15}$
8	16
9	$3\sqrt{3}$
10	$6\sqrt{5}$
11	$4\sqrt{15}$
12	x
13	$x\sqrt{x}$
14	k^6
15	x^3
16	$x^2\sqrt{x}$
17	x^4
18	$x^{\frac{12}{5}}$
19	$7x\sqrt{2y}$
20	$x = 4$
21	$a = 9$
22	$\sqrt[3]{x^2}$
23	$x^{\frac{11}{6}}$ or $\sqrt[6]{x^{11}}$
24	$2\sqrt[3]{z}$
25	$3ab^2$
26	$y^2 z\sqrt[3]{x^2 z}$
27	$\dfrac{\sqrt{6}}{3}$
28	$2 + \dfrac{2x}{3}$
29	$x = \text{-}2, 1$
30	$12\sqrt{3x}$

Radicals
Algebra and Arithmetic Problem Set 18

$\sqrt{18} =$

A. 9

B. $2\sqrt{3}$

C. $3\sqrt{2}$

D. $9\sqrt{2}$

E. 324

$\sqrt{729} =$

F. 27

G. $3\sqrt{81}$

H. $9\sqrt{9}$

J. $\sqrt{27}$

K. 531,441

$\sqrt[3]{27} =$

A. 9

B. 3

C. 19,683

D. $3\sqrt{3}$

E. $3\left(\sqrt[3]{3}\right)$

Which of the following is equal to $9^{\frac{1}{9}}$?

F. 1

G. 0

H. $3\left(\sqrt[3]{9}\right)$

J. $\sqrt[9]{9}$

K. 9^{-9}

DO YOUR FIGURING HERE

Radicals
Algebra and Arithmetic Problem Set 18

5. Which of the following is equal to $8^{\frac{1}{3}}$?

   A. $\sqrt[3]{8}$

   B. $2\left(\sqrt[3]{8}\right)$

   C. 8^{-3}

   D. $\dfrac{8}{3}$

   E. $\dfrac{3}{8}$

DO YOUR FIGURING HERE

6. $\dfrac{8}{\sqrt{5}} + \dfrac{7}{\sqrt{3}} =$

   F. $\dfrac{8\sqrt{3}+7\sqrt{5}}{\sqrt{8}}$

   G. $\dfrac{8\sqrt{3}+7\sqrt{5}}{\sqrt{15}}$

   H. $\dfrac{15}{\sqrt{15}}$

   J. $\dfrac{15}{\sqrt{8}}$

   K. $\dfrac{15}{\sqrt{5}+\sqrt{3}}$

7. What are the real number values of x that make the equation $\sqrt[4]{x^{24}} = x^6$ true?

   A. All Real Numbers

   B. $x < 0$

   C. $x > 0$

   D. $x \leq 0$

   E. $x \geq 0$

8. Find one value of x
$$2\sqrt{4x^4} = x^4$$

   F. -2

   G. 4

   H. 8

   J. 12

   K. 16

Radicals
Algebra and Arithmetic Problem Set 18

Solve for x

$\sqrt[3]{x} + 2\sqrt[3]{x} = 3x$

A. $\dfrac{1}{9}$

B. 1

C. 3

D. 6

E. 9

DO YOUR FIGURING HERE

$\sqrt[3]{243} =$

F. 9

G. $3\left(\sqrt[3]{3}\right)$

H. $3\left(\sqrt[3]{9}\right)$

J. 27

K. $9\left(\sqrt[3]{9}\right)$

$\sqrt{98x^5} =$

A. $24x^4\sqrt{2}$

B. $7x^2$

C. $7x^2\sqrt{2x}$

D. $49x^2\sqrt{2x}$

E. $98x\sqrt{2}$

Solve for x:

$\sqrt{3x-6} - 2\sqrt{x+7} = 0$

F. -34

G. 0

H. 20

J. 34

K. No real number

2019-1LogicPrep ALGEBRA 183

Radicals
Algebra and Arithmetic Problem Set 18

13. $\left(\dfrac{x^2y^3}{y^2}\right)^{\frac{1}{2}} =$

DO YOUR FIGURING HERE

A. $xy^{-\frac{1}{2}}$

B. $\dfrac{1}{x^2y}$

C. $x\sqrt{y}$

D. $xy\sqrt{x}$

E. $\dfrac{x^2y}{2}$

14. Solve for x:

$3\,(x)^{-\frac{1}{2}} = 4$

F. $x = -\dfrac{9}{16}$

G. $x = \dfrac{9}{16}$

H. $x = \dfrac{2}{3}$

J. $x = \dfrac{3}{4}$

K. $x = \dfrac{16}{9}$

15. Solve for x:

$\left(\dfrac{9}{4}\right)^{\frac{1}{2}} - \left(\dfrac{\sqrt{x}}{\sqrt{7}}\right)^{-2} = 3x^{-1}$

A. $x = \dfrac{21}{44}$

B. $x = \dfrac{14}{22}$

C. $x = \dfrac{44}{9}$

D. $x = \dfrac{20}{3}$

E. $x = \dfrac{6+2\sqrt{7}}{9}$

ALGEBRA

What is a possible value of x?

$$\sqrt{x} + \frac{1}{3\sqrt{x}} = 3\sqrt{x}$$

F. $-\frac{1}{6}$

G. $\frac{1}{36}$

H. $\frac{1}{6\sqrt{2}}$

J. $\frac{1}{6}$

K. $\frac{1}{9}$

DO YOUR FIGURING HERE

Radicals
Algebra and Arithmetic Problem Set 18

Answer Key

#	Answer	Frequency	Difficulty
1	C	popular	1
2	F	popular	1
3	B	popular	1
4	J	popular	2
5	A	popular	2
6	G	popular	2
7	A	popular	3
8	F	popular	2
9	B	popular	1
10	H	popular	3
11	C	popular	2
12	K	popular	3
13	C	popular	2
14	G	popular	2
15	D	popular	3
16	J	popular	3

Inequalities
Quick Drill

1. Solve for x:
   $2x - 3 < 1$

2. Solve for x:
   $x - 4 > 2$

3. Solve for m:
   $3m \leq 3$

4. Solve for b:
   $-b - 8 > 2$

5. If $-2 < x < 6$ then what is the range of values of $3x + 4$?

6. If $x^2 < 4$ and $y > 3$ then what is the range of values of $\dfrac{x}{y}$?

7. Solve for f:
   $-\dfrac{f}{3} + 4 \leq 16$

8. If $-3 \leq m \leq 3$ and $0 \leq n \leq 5$ then what is the range of values of mn?

9. Solve for b:
   $\dfrac{9 - 3b}{2} < b - 3$

10. Solve for z:
    $-3z + 27 \geq 36$

Inequalities
Quick Drill

Answer Key

#	Answer
1	$x < 2$
2	$x > 6$
3	$m \leq 1$
4	$b < -10$
5	$-2 < 3x + 4 < 22$
6	$-\dfrac{2}{3} < \dfrac{x}{y} < \dfrac{2}{3}$
7	$f \geq -36$
8	$-15 \leq mn \leq 15$
9	$b > 3$
10	$z \leq -3$

Absolute Value
Quick Drill

1. $|7 - 3| - |\text{-}2 + 1| =$

2. $\text{-}|4 - 8| =$

3. $(\text{-}|3 - 6|)^2 =$

4. $8 - |9 - 20| =$

5. $|8 - 9 - 20| =$

6. Solve for x:
   $|x - 2| = 3$

7. Solve for x:
   $|3x| > 9$

8. Solve for x:
   $|3x - 4| \le 1$

9. Solve for x:
   $|4x + 2| < 0$

10. Solve for x:
    $|7x + 1| > 3$

11. Solve for a:
    $|a + 1| \le 6$

Absolute Value
Quick Drill

12. Solve for d:
$$|3d - 4| = 10$$

17. Solve for g:
$$3 \geq |\frac{2g + 1}{3}|$$

13. Solve for b:
$$-3|b + 4| = 8$$

18. Solve for c:
$$|-7c - 6| > 12$$

14. If $|3 - 2m| > 16$, what are the possible values of m?

19. Solve for b:
$$|-8| > |2b + 2|$$

15. What is the solution set to $|x - 8| \leq 8$?

20. Solve for k:
$$k|-3| \geq |k + 18|$$

16. Solve for x:
$$|16x + 3| \geq 19$$

Absolute Value
Quick Drill

Answer Key

#	Answer
1	3
2	-4
3	9
4	-3
5	-21
6	$x = -1, 5$
7	$x < -3$ or $x > 3$
8	$1 \leq x \leq \dfrac{5}{3}$
9	No solution
10	$x < -\dfrac{4}{7}$ or $x > \dfrac{2}{7}$
11	$-7 \leq a \leq 5$
12	$d = -2, \dfrac{14}{3}$
13	No solution
14	$m < -\dfrac{13}{2}$ or $m > \dfrac{19}{2}$
15	$0 \leq x \leq 16$
16	$x \leq -\dfrac{11}{8}$ or $x \geq 1$
17	$-5 \leq g \leq 4$
18	$c < -\dfrac{18}{7}$ or $c > \dfrac{6}{7}$
19	$-5 < b < 3$
20	$k \geq -\dfrac{9}{2}$

Absolute Value
Algebra and Arithmetic Problem Set 19

1. $|9 - 4| - |3 - 7| = ?$

    A. -9

    B. -1

    C. 1

    D. 9

    E. 23

DO YOUR FIGURING HERE

2. What is the value of $|5 - x|$ if $x = 8$?

    F. -3

    G. 3

    H. 8

    J. 13

    K. 26

3. If $|x + 2| > 3$, then what is one possible value of x?

    A. -3

    B. -1

    C. 0

    D. 1

    E. 2

4. $f(x) = 4x - |3x - 5|$. What is $f(1)$?

    F. 1

    G. 2

    H. 4

    J. 5

    K. 6

5. Which number is NOT included in the solution set of $|2x + 5| \geq 1$?

    A. -4.25

    B. -2.50

    C. -1.00

    D. 4.25

    E. Solution set is all real numbers

Absolute Value
Algebra and Arithmetic Problem Set 19

If $|3 - 2m| > 16$, which of the following is a possible value for m?

F. -7

G. -5

H. 0

J. 3

K. 9

DO YOUR FIGURING HERE

$f(x) = x + 3$ and $g(x) = |x + 3|$. When are these two functions the same?

A. For no values of x

B. $x \geq -3$

C. $x \geq 0$

D. $x \geq 3$

E. For all values of x

Roger tries to keep his speed on the highway between 5 mph above and 5 mph below the speed limit. Which inequality expresses his desired speed on a road with a 55 mph limit?

F. $s - 55 > 5$

G. $s - 5 \leq 55$

H. $|s - 55| \leq 5$

J. $|s - 55| = 5$

K. $|s - 55| \geq 5$

What is the solution set of $|2x + 2| \geq 6$?

A. $\{x : x \geq -4\}$

B. $\{x : x \geq 2\}$

C. $\{x : x \leq -4 \text{ or } x \geq 2\}$

D. $x : x \leq -8 \text{ or } x \geq 4\}$

E. $\{\quad\}$ (the empty set)

10. The graph of $y = f(x)$ is shown in the standard (x, y) coordinate plane below. Which of the following graphs is that of $y = |f(x)|$?

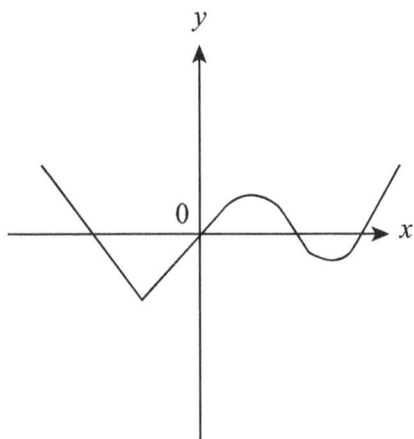

DO YOUR FIGURING HERE

F.

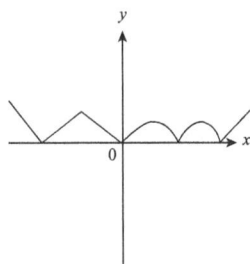

G.

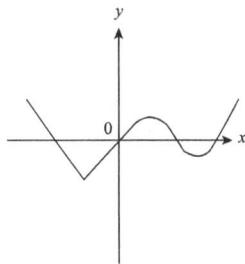

H.

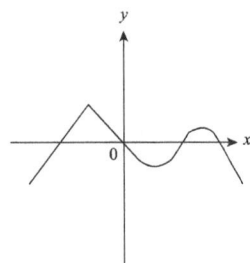

J.

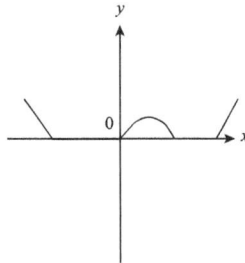

K.

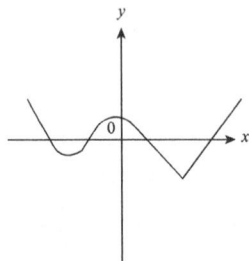

Absolute Value
Algebra and Arithmetic Problem Set 19

Molly turns on the heat if it is 58°F or colder. She turns on a fan if it is 82°F or hotter. Which of the following inequalities shows the temperatures, t, at which neither the heat nor the fan is on?

A. $t < 82$

B. $|t - 70| < 12$

C. $|t - 70| > 24$

D. $|t - 58| < 24$

E. $|t - 70| > 12$

There is a system of inequalities:
$$-3x - 2 \text{ if } x \leq -3$$
$$3x + 2 \text{ if } x \geq \frac{5}{3}$$
Which of the following inequalities has the same solution set?

F. $-3x + 2 \leq 3$

G. $3x + 2 \leq -3$

H. $|-3x - 2| \leq -3$

J. $|3x + 2| \geq \frac{5}{3}$

K. $|-3x - 2| \geq 7$

For real numbers n and m such that $-1 < \dfrac{m}{n} < 1$, where m and n are not equal to 0, what must be the relationship between m and n?

A. m and n are equal.

B. One is positive and the other is negative.

C. $|m| < |n|$

D. $|n| < |m|$

E. m and n are both negative.

Absolute Value
Algebra and Arithmetic Problem Set 19

14. On a real number line, how far apart are the two solutions to the equation $|x - n| = 9$, for any real number n?

    F. n

    G. $2n$

    H. $n + 9$

    J. $2(n + 9)$

    K. 18

DO YOUR FIGURING HERE

15. A regulation for riding the bumper cars at a certain amusement park requires that all children be between 36 and 60 inches tall. Which of the following inequalities can be used to determine whether or not a child's height, h, satisfies the rules for this ride?

    A. $|h - 60| < 8$

    B. $|h - 48| < 12$

    C. $|h - 48| > 36$

    D. $|h - 36| < 24$

    E. $|h - 36| < 60$

16. For bowling, Mr. Torres will not use any bowling balls that weigh less than 8 pounds or more than 14 pounds. If x represents the weight of a bowling ball, in pounds, that Mr. Torres will not use, which of the following inequalities represents all possible values of x?

    F. $|x - 8| < 14$

    G. $|x - 8| < 6$

    H. $|x - 10| < 4$

    J. $|x - 11| < 3$

    K. $|x - 14| < 8$

17.
$$f(x) = |\, 2x - 13 \,|$$
For the function above, what is one possible value of x for which $f(x) < x$

Absolute Value
Algebra and Arithmetic Problem Set 19

$$|x + 4| = \frac{1}{2}$$

What is the greatest value of x thats satisfies the equation above?

DO YOUR FIGURING HERE

Absolute Value
Algebra and Arithmetic Problem Set 19

Answer Key

#	Answer	Frequency	Difficulty
1	C	popular	1
2	G	popular	1
3	E	popular	2
4	G	popular	1
5	B	popular	2
6	F	popular	3
7	B	popular	2
8	H	popular	2
9	C	popular	2
10	F	average	1
11	B	popular	3
12	K	popular	4
13	C	popular	3
14	K	popular	3
15	B	popular	4
16	J	popular	4
17	$\frac{13}{3} < x < 13$	popular	5
18	-3.5 or $\frac{-7}{2}$	popular	2

Functions
Quick Drill

1. $f(x) = 3x^2 + 2x - 3$. What is $f(2)$?

2. $g(a) = \dfrac{a}{2} + 8$. What is $g(8)$?

3. $f(r) = \dfrac{r^2 + r + 3}{r}$. What is $f(3)$?

4. $g(x) = x^3 + 2x^2 - 5$. What is $g(2)$?

5. $h(m) = (3m)^2$. What is $h(4)$?

6. $f(b) = 4b - \dfrac{3}{2}$. When $f(b) = \dfrac{2}{3}$, what is b?

7. $g(y) = y^2 - 2y$. When $g(y) = 0$, what is y?

8. $f(x) = 3x - 4$. When $f(x) = 6$, what is x?

9. $h(s) = \dfrac{s}{3} + 14$. When $h(s) = 7$, what is s?

10. $f(n) = 2n - 6$. When $3(f(n)) = 15$, what is n?

Functions
Quick Drill

11. $f@g = -3fg + 6g$. What is $2@\text{-}2$?

12. $a\#b = \dfrac{a}{b} - \dfrac{b}{a}$. What is $3\#12$?

13. $k\star = k^2 - 4k$. What is $\dfrac{1}{2\star}$?

14. $x\%y = 16xy + 3y$. What is $2\%\,(1\%5)$?

15. $r;s = r^s - r$. What is $7;2$?

16. $a?b = 4ab + a - 2b$. What is $(9?3)\,?1$?

17. $g!h = 3g + h^2$. What is $2!\,(5!3)$?

18. $x"y"z = xy + xz + yz$. What is $6"3"2$?

19. $\star h = 18 - 2h$. When $\star h = 14$, what is h?

20. $a, = 3\,(a - 2)$. When $a, = 22$, what is a?

21. $f(x) = 3x$ and $g(x) = x^2$. What is $f(g(3))$?

Functions
Quick Drill

22. $f(a) = a - 2$ and $g(a) = 6a + 2$. What is $g(f(2))$?

23. $f(r) = 3r^2 + 4r + 1$ and $g(r) = r - 8$. What is $f(g(4))$?

24. $f(b) = b(b + 4)$ and $g(b) = \dfrac{b}{3}$. What is $f(3(g(4)))$?

25. $f(x) = 3x - 14$ and $g(x) = \dfrac{x + 2}{4}$. For what value of x does $f(x) = g(x)$?

26. $f(a) = \dfrac{a}{2}$ and $g(a) = \dfrac{a + 4}{3}$. For what value of a does $f(a) = g(a)$?

27. $f(b) = 6b - \dfrac{b}{4}$ and $g(b) = 2b + 1$. For what value of b does $f(b) = g(b)$?

28. $g(x) = x^2 - 1$
$t(x) = g(4x) + 5$
$t(a) = 260$
What is the value of $g(a)$?

29. $g(x) = 5x + 3$
$p(x) = g(4x) + 8$
$g(b) = 198$
What is the value of $p(b)$?

30. $g(x) = x^2 + 1$
$f(x) = g(3x) - 4$
$f(a) = 897$
What is the value of a?

Answer Key

#	Answer
1	13
2	12
3	5
4	11
5	144
6	$b = \dfrac{13}{24}$
7	$y = 0, 2$
8	$x = \dfrac{10}{3}$
9	$s = -21$
10	$n = \dfrac{11}{2}$
11	0
12	$-\dfrac{15}{4}$
13	$-\dfrac{1}{4}$
14	3325
15	42
16	553
17	582
18	36
19	$h = 2$
20	$a = \dfrac{28}{3}$
21	27
22	2
23	33
24	32
25	$x = \dfrac{58}{11}$
26	$a = 8$
27	$b = \dfrac{4}{15}$
28	15
29	791
30	± 10

Functions I
Algebra and Arithmetic Problem Set 20

1. Let $f(x) = \dfrac{x^2 + 8}{x - 4}$. What is the value of $f(8)$?

   A. 72

   B. 18

   C. 8

   D. 4

   E. 2

2. If $f(x) = 7x^2 + 3x - 4$, then $f(-2) = ?$

   F. -38

   G. -26

   H. 18

   J. 30

   K. 39

3. If $n(x) < 0$ for all real values of x, which of the following could be the function n?

   A. $n(x) = 2x + 1$

   B. $n(x) = -x$

   C. $n(x) = -2$

   D. $n(x) = x^2 - 8$

   E. $n(x) = x^3$

4. What values of x satisfy the equation $(x + a)(x + b) = 0$?

   F. $-a$ and $-b$

   G. $-a$ and b

   H. $-ab$

   J. a and $-b$

   K. a and b

5. If $f(x) = x^2 + 3x - 4$ and $g(x) = 2\sqrt{x}$, then what is the value of $\dfrac{g(9)}{f(2)}$?

A. 1

B. $\dfrac{52}{3}$

C. $\dfrac{3}{52}$

D. 2

E. $\dfrac{\sqrt{2}}{2}$

DO YOUR FIGURING HERE

6. When $(3x - 5)^2$ is written in the form $ax^2 + bx + c$, where a, b, and c are integers, $a + b + c =?$

F. -46

G. -16

H. 4

J. 19

K. 64

7. What is the sum of the 2 solutions of the equation $x^2 + 2x - 24 = 0$?

A. -10

B. -2

C. 0

D. 1

E. 2

Functions I
Algebra and Arithmetic Problem Set 20

DO YOUR FIGURING HERE

8. The quadratic formula gives the 2 roots
$x = \dfrac{-b \pm \sqrt{b^2 - 4ac}}{2a}$ for the equation
$ax^2 + bx + c = 0$. What are the 2 roots for the
equation $x^2 - x = 12$?

  F. $\dfrac{1 \pm \sqrt{-13}}{2}$

  G. $\dfrac{1 \pm \sqrt{13}}{2}$

  H. -4 and 3

  J. -3 and 4

  K. 2 and $\dfrac{3}{2}$

9. For a certain quadratic equation,
$ax^2 + bx + c = 0$, the two solutions are $x = \dfrac{5}{3}$ and
$x = -2$. Which of the following could be factors of
$ax^2 + bx + c = 0$?

  A. $(3x - 5)$ and $(4x - 8)$

  B. $(3x - 5)$ and $(4x + 8)$

  C. $(3x + 5)$ and $(4x - 8)$

  D. $(3x - 5)$ and $(4x - 2)$

  E. $(3x - 5)$ and $(4x + 2)$

10. What is the range of the function $g(x) = |x + 8|$?

  F. $x > -8$

  G. $x > 0$

  H. $g(x) > -8$

  J. $g(x) > 0$

  K. $g(x) > 8$

11. What is the range of the function defined as
$f(x) = x^2 - 3$?

  A. All real numbers

  B. All integers

  C. $f(x) \geq 0$

  D. $g(x) \geq -3$

  E. $g(x) \geq 3$

Functions I
Algebra and Arithmetic Problem Set 20

12. In the xy-coordinate plane below, the quadrants are labeled I through IV. Line m (not shown) does not contain points in either quadrant I or quadrant III. Which of the following could be the equation of line m?

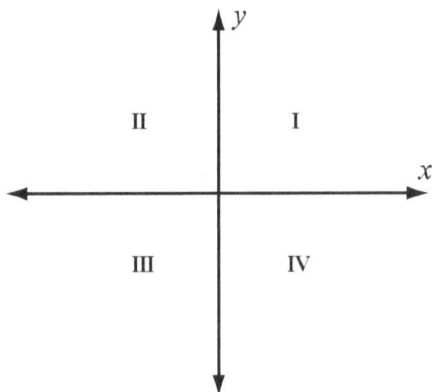

F. $y = 1$

G. $x = 1$

H. $y = -3x$

J. $y = -3x - 5$

K. $y = 3x - 5$

13. If $f(x) = x^2 - 6$, then $f(x + h) = ?$

A. $x^2 + h^2$

B. $x^2 - 6 + h$

C. $x^2 + h^2 - 6$

D. $x^2 + 2xh + h^2$

E. $x^2 + 2xh + h^2 - 6$

14. If $f(x) = x^2 + 1$ and $g(x) = 2x + 1$, what is $f(g(3)) = ?$

F. 20

G. 21

H. 49

J. 50

K. 51

Functions I
Algebra and Arithmetic Problem Set 20

- If $f(x) = x^2 + 1$ and $g(x) = 2x + 1$, what is $g(f(x+1))$?

  A. $2x^2 + 3$

  B. $2x^2 + 4x + 5$

  C. $4x^2 + 12x + 10$

  D. $4x^2 + 4x + 2$

  E. $x^2 + 4x + 1$

- Given $h(x) = x + \left(\dfrac{2}{x}\right)$ and $j(x) = \dfrac{1}{x}$, what is the value of $j(h(1))$

  F. $\dfrac{1}{3}$

  G. 1

  H. $\dfrac{3}{2}$

  J. 3

  K. 6

- If $(x - 7)$ is a factor of $3x^2 - 19x + k$, what is the value of k?

  A. -21

  B. -14

  C. -7

  D. 7

  E. 28

- Let @ be defined by $a@b = ab + a - b$ for all numbers a and b. $2@5 = ?$

  F. 4

  G. 7

  H. 9

  J. 11

  K. 13

Functions I
Algebra and Arithmetic Problem Set 20

19. If * represents the operation defined by
$a * b = a^b + b$, then $4 * (3 * 1) =$?

    A. 64

    B. 67

    C. 68

    D. 256

    E. 260

DO YOUR FIGURING HERE

20. Let @ be defined by $a@b = ab + a - b$ for all
numbers a and b. If $10@n = 73, n = $?

    F. 4

    G. 5

    H. 6

    J. 7

    K. 8

21. For all values of x, let $x\$$ be defined by
$x\$ = x^2 + 1$. Which of the following is equivalent
to $(x\$) \$$?

    A. $x^4 + 2$

    B. $x^4 + 2x$

    C. $x^4 + 2x^2$

    D. $x^4 + 2x^2 + 1$

    E. $x^4 + 2x^2 + 2$

22. $a@b = \dfrac{a}{b} - \dfrac{b}{a}$. When $a > b$, which of the
following is true?

    F. $a@b < b@a$

    G. $a@b = b@a$

    H. $a@b > b@a$

    J. $a@b = a(b@a)$

    K. $a@b = b(b@a)$

Functions I
Algebra and Arithmetic Problem Set 20

23. Which of the following quadratic equations has solutions $x = 4a$ and $x = -7b$

   A. $x^2 - 28ab = 0$

   B. $x^2 - x(7b - 4a) - 28ab = 0$

   C. $x^2 - x(7b + 4a) + 28ab = 0$

   D. $x^2 + x(7b - 4a) - 28ab = 0$

   E. $x^2 + x(7b + 4a) + 28ab = 0$

DO YOUR FIGURING HERE

24. Which of the following is the graph of the function $f(x)$ defined below?

$$f(x) = \begin{cases} x^2 - 3 & \text{for } x \leq -2 \\ x - 4 & \text{for } -2 < x < 4 \\ 5 - x & \text{for } x \geq 4 \end{cases}$$

G.

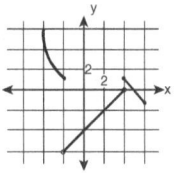
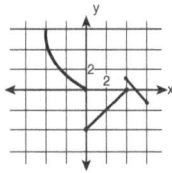

J.

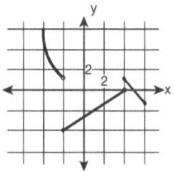
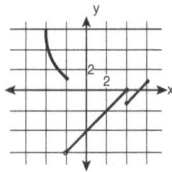

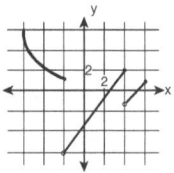

25. Let the function $f(a, b)$ be defined as $f(a, b) = b^2 - a$. For all x and y, $f\left((x^2 + y^2), (x - y)\right) = ?$

   A. $2y^2$

   B. 0

   C. $-2y^2$

   D. $-2xy + 2y^2$

   E. $-2xy$

Functions I
Algebra and Arithmetic Problem Set 20

Answer Key

#	Answer	Frequency	Difficulty
1	B	popular	1
2	H	popular	1
3	C	rare	2
4	F	popular	1
5	A	popular	1
6	H	popular	1
7	B	popular	2
8	J	popular	2
9	B	popular	2
10	J	popular	1
11	D	popular	2
12	H	popular	3
13	E	popular	1
14	J	average	1
15	B	average	3
16	F	average	1
17	B	popular	2
18	G	average	2
19	E	popular	2
20	J	average	2
21	E	average	2
22	H	average	3
23	D	popular	2
24	F	popular	2
25	E	popular	2

Functions II
Algebra and Arithmetic Problem Set 21

An equation $7x + 4y = 9$ is graphed on a coordinate plane. What is the slope of this line?

A. $\dfrac{-9}{4}$

B. $\dfrac{-7}{4}$

C. $\dfrac{-7}{9}$

D. $\dfrac{7}{4}$

E. $\dfrac{9}{4}$

DO YOUR FIGURING HERE

What does y equal when $5x + 3y - 17 = 0$?

F. $\dfrac{-3x + 17}{5}$

G. $\dfrac{-5x + 17}{3}$

H. $\dfrac{5x + 17}{3}$

J. $-5x + 17$

K. $-3x + 17$

How many ordered pairs satisfy the equation $3x - 7y = 9$?

A. 0

B. 1

C. 2

D. 3

E. infinitely many

4.

$$y = \frac{1}{2}(x+3)^2 - 4$$

In the xy-plane, line n passes through the point $(3, 14)$ and the vertex of the parabola determined by equation above. What is the slope of line n?

F. 0

G. $\dfrac{1}{3}$

H. 1

J. 3

K. 6

5. For all $x \neq$ -6, $\quad \dfrac{6x + x^2}{x^2 + 4x - 12} =?$

A. $\dfrac{x}{x+6}$

B. $\dfrac{-x}{x-2}$

C. $\dfrac{x}{x-2}$

D. $\dfrac{-1}{2}$

E. $\dfrac{1}{6}$

6. What is $f(x)$, given that $f^{-1}(x) = 3x - 18$?

F. $f(x) = \dfrac{x+3}{18}$

G. $f(x) = \dfrac{x+18}{3}$

H. $f(x) = \dfrac{3x+18}{2}$

J. $f(x) = 18x + 3$

K. $f(x) = x + 6$

DO YOUR FIGURING HERE

Functions II
Algebra and Arithmetic Problem Set 21

7. Which of the following is the inverse of
$f(x) = x^2 - 3$?

   A. $f^{-1}(x) = 3x$

   B. $f^{-1}(x) = \sqrt{3x}$

   C. $f^{-1}(x) = \pm\sqrt{x+3}$

   D. $f^{-1}(x) = 3\sqrt{x}$

   E. $f^{-1}(x) = -3\sqrt{x}$

8. Which of the following is the inverse of
$f(x) = \dfrac{2x - 3}{x + 1}$?

   F. $f^{-1}(x) = \dfrac{x + 3}{2 - x}$

   G. $f^{-1}(x) = \dfrac{x - 2}{3 - x}$

   H. $f^{-1}(x) = \dfrac{x + 1}{2x - 3}$

   J. $f^{-1}(x) = \dfrac{x - 1}{2x + 3}$

   K. $f^{-1}(x) = (x - 1)(2x + 3)$

9. Which of the following is the inverse of
$f(x) = \dfrac{x^2 + 1}{x^2 - 1}$?

   A. $f^{-1}(x) = \pm\sqrt{\dfrac{x + 1}{x - 1}}$

   B. $f^{-1}(x) = f(x)$

   C. $f^{-1}(x) = \dfrac{\pm\sqrt{x + 1}}{x - 1}$

   D. $f^{-1}(x) = \dfrac{\sqrt{x + 1}}{x - 1}$

   E. $f^{-1}(x) = \dfrac{x + 1}{x - 1}$

DO YOUR FIGURING HERE

LogicPrep ALGEBRA 219

10. How many times does the graph of
$y = (x + 2)(x - 2)(x + 7)(x + 5)$ intersect the x-axis?

F. 1

G. 2

H. 3

J. 4

K. 5

DO YOUR FIGURING HERE

11. If 5 is a solution to the equation
$x^2 - hx + 10 = 0$, then what does h equal?

A. -7

B. -5

C. -2

D. 2

E. 7

12. What is the y-intercept of the graph of
$4x + 5y = 13$?

F. $\dfrac{4}{5}$

G. $\dfrac{5}{13}$

H. $\dfrac{13}{5}$

J. $\dfrac{5}{4}$

K. $\dfrac{4}{13}$

Functions II
Algebra and Arithmetic Problem Set 21

Which of the following is the inverse of
$f(x) = 3x^2 + 2$?

A. $f^{-1}(x) = \dfrac{\pm\sqrt{x} - 2}{3}$

B. $f^{-1}(x) = \dfrac{x \pm \sqrt{2}}{3}$

C. $f^{-1}(x) = \pm\sqrt{\dfrac{x - 2}{3}}$

D. $f^{-1}(x) = \pm\dfrac{\sqrt{3x - 6}}{2}$

E. $f^{-1}(x) = 3x - 6$

DO YOUR FIGURING HERE

If $f(x) = 2x + 14$, then what is the value of
$f^{-1}(2)$?

F. -16

G. -13

H. -6

J. 8

K. 18

The function $f(x) = \dfrac{x + 2}{x^2 - x - 6}$ is
discontinuous at how many values of x?

A. 0

B. 1

C. 2

D. 3

E. 4

The function $f(x) = \dfrac{2x + 2}{x^2 - 1}$ has a vertical
asymptote at

F. $x = -1$

G. $x = 0$

H. $x = 1$

J. $x = 2$

K. there is no vertical asymptote

17. When $a \neq 0$ and $b \neq 0$, the graph of

$f(x) = \dfrac{3x + b}{2x + a}$ has a horizontal asymptote at

A. $y = \dfrac{3}{2}$

B. $y = \dfrac{2}{3}$

C. $y = \dfrac{b}{2}$

D. $y = \dfrac{3}{a}$

E. $y = b$

DO YOUR FIGURING HERE

Functions II
Algebra and Arithmetic Problem Set 21

nswer Key

#	Answer	Frequency	Difficulty
1	B	popular	1
2	G	popular	1
3	E	average	2
4	J	rare	2
5	C	rare	3
6	G	rare	2
7	C	rare	2
8	F	rare	2
9	A	rare	2
10	J	average	1
11	E	popular	3
12	H	average	1
13	C	rare	2
14	H	rare	2
15	C	rare	4
16	H	rare	4
17	A	rare	4

Interest
Algebra and Arithmetic Problem Set 22

1. A formula used to compute the current value of a savings account is $A = P(1 + r)^n$, where A is the current value; P is the amount deposited; r is the rate of interest for 1 compounding period, expressed as a decimal; and n is the number of compounding periods. Which of the following is closest to the value of a savings account after 8 years if $\$15,000$ is deposited at 6% annual interest compounded yearly?

   A. $\$15,735.30$

   B. $\$15,900.00$

   C. $\$23,803.11$

   D. $\$23,907.72$

   E. $\$27,763.95$

2. A painting is purchased for 20,000 dollars. The value of the painting can be estimated using a compound interest at an annual rate of 26%. How much is it worth after 12 years?

   F. $\$320,000.00$

   G. $\$320,240.71$

   H. $\$300,000.70$

   J. $\$320,240.00$

   K. $\$452,927.59$

3. The value of a VW Jetta decreases each year by 12 percent and Lindsay paid $\$25,000$ for a new VW Jetta. If the value v of the car x years from when she first purchased the car is given by the function $v(x) = 25,000 \, (p)^x$, what is the value of p?

   A. .12

   B. .88

   C. 1.12

   D. 1.82

   E. 1.88

DO YOUR FIGURING HERE

Interest
Algebra and Arithmetic Problem Set 22

Kavitha invested $1,500 on January 1. At the end of 6 months, during which time Kavitha made no withdrawals and no deposits, the investment has earned $80 in interest. Kavitha's $1,500 investment returned an annual percentage yield closest to which of the following percents?

F. 8%

G. 9%

H. 10%

J. 11%

K. 12%

At a clothing store, items were put on a rail and assigned prices. Each week after, the price was 15 percent less than the price for the previous week. If the price of an item was d dollars for week 1, what was the price for week 4?

A. $.4d$

B. $.522d$

C. $.61413d$

D. $.7225d$

E. $.85d$

$$b(t) = 1,000\,(.73)^t$$

The function above can be used to model the population of a certain bee colony. If $b(t)$ gives the number of the bees living t decades after the year 1980, which of the following is true about the population of the bee colony from 1980 to 2010?

F. It decreased by about 390

G. It increased by about 390

H. It decreased by about 610

J. It increased by about 610

K. It remained the same

DO YOUR FIGURING HERE

7. Depreciation can be modeled by the formula $V = I(1 - r)^t$, where I is the cars initial purchase price, r is the car's constant annual rate of decrease in value, expressed as a decimal: and V is the car's dollar value at the end of t years.

A used car with a purchase price of $60,000 has a constant annual rate of decrease in value of 0.3. According to the model, what is the value of the car, to the nearest dollar, at the end of 4 years?

A. $8,076

B. $10,085

C. $14,406

D. $20,580

E. $24,576

8. Depreciation can be modeled by the formula $V = I(1 - r)^t$, where I is the cars initial purchase price, r is the car's constant annual rate of decrease in value, expressed as a decimal: and V is the car's dollar value at the end of t years.

A used car decreased in value by 14% over 2 years. The car's initial price was $20,000. At what rate did the value of the car decrease, to the nearest hundredth?

F. 0.057

G. 0.073

H. 0.078

J. .860

K. .927

DO YOUR FIGURING HERE

Interest
Algebra and Arithmetic Problem Set 22

Answer Key

#	Answer	Frequency	Difficulty
1	D	popular	1
2	G	popular	2
3	B	popular	2
4	J	popular	2
5	C	popular	3
6	H	popular	3
7	C	popular	3
8	G	popular	3

Venn Diagrams, Unions, and Intersections
Algebra and Arithmetic Problem Set 23

1. The Venn diagram below shows the distribution of Ms. Harper's kindergarten class' animal preference, showing how many students liked rabbits, how many liked reindeer, how many liked both and how many liked neither. How many liked reindeer?

DO YOUR FIGURING HERE

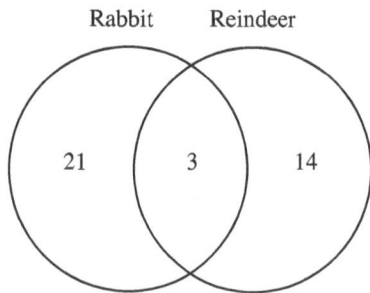

Rabbit   Reindeer

```
21    3    14
```

A. 3

B. 14

C. 17

D. 18

E. 21

2. The Venn diagram below shows how many items are in categories A, B and C How many items are in both categories A and C?

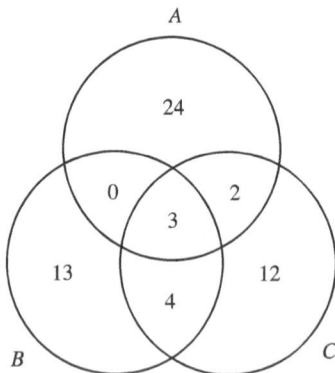

A

```
        24
     0      2
        3
  13          12
        4
B              C
```

F. 2

G. 3

H. 5

J. 36

K. 41

Venn Diagrams, Unions, and Intersections
Algebra and Arithmetic Problem Set 23

3. How many items are in category A in the diagram below?

A. 5

B. 24

C. 26

D. 27

E. 29

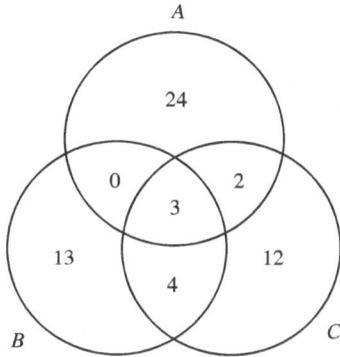

DO YOUR FIGURING HERE

4. The diagram below illustrates the features of cars. What features do the cars in the section marked x have?

F. Four-wheel drive and GPS but NOT anti-lock brakes.

G. Four-wheel drive, GPS and anti-lock brakes.

H. Four-wheel drive only.

J. GPS only.

K. Anti-lock brakes and GPS

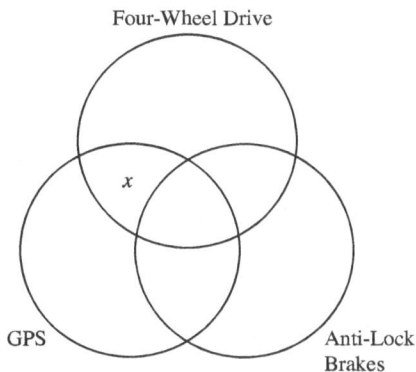

Venn Diagrams, Unions, and Intersections
Algebra and Arithmetic Problem Set 23

5. The Venn diagram below shows the distribution of languages spoken by a group of students. How many students speak exactly two languages?

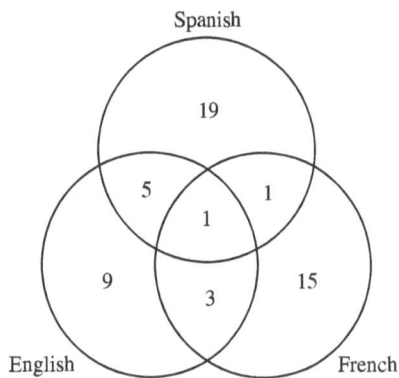

A. 1

B. 3

C. 5

D. 8

E. 9

6. Set A $\{4, 8, -3, 9, 8, 6, 12\}$
    Set B $\{3, 9, 6, -3, 3, 5, 7\}$
    List the union of Sets A and B.

7. Set A $\{4, 8, -3, 9, 8, 6, 12\}$
    Set B $\{3, 9, 6, -3, 3, 5, 7\}$
    What is the intersection of Sets A and B?

8. Set X $\{-2, 4, 12, 8, 4, 10, 12\}$
    Set Y $\{-7, -2, 3, 4, -7, 6, 4\}$
    How many integers are in the union of Set X and Set Y?

9. Set X $\{-2, 4, 12, 8, 4, 10, 12\}$
    Set Y $\{-7, -2, 3, 4, -7, 6, 4\}$
    List the intersection of Set X and Set Y.

Venn Diagrams, Unions, and Intersections
Algebra and Arithmetic Problem Set 23

0. In the figure below, which of the labeled points is inside circle A and circle B but not inside circle C?

F. M

G. N

H. O

J. P

K. Q

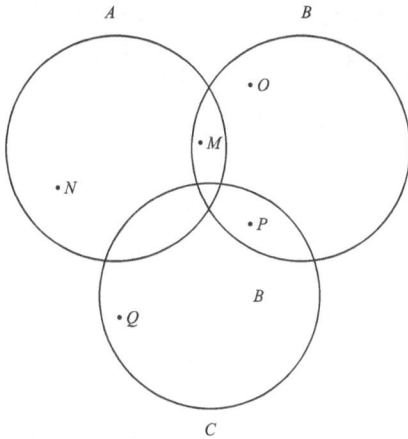

1. If A is the set of prime numbers between 1 and 15 and B is the set of all positive multiples of 2 or 3, how many numbers appear in both A and B?

A. One

B. Two

C. Three

D. Four

E. Five

2. In a high school, 240 students take Spanish and 180 students take French. If there are a total of 310 students in the school and every student must take French, Spanish or both, then how many of the students take both French and Spanish?

Venn Diagrams, Unions, and Intersections
Algebra and Arithmetic Problem Set 23

13. There are 26 kids in a class. If 16 students have brown hair, 14 students have blue eyes, and there are no kids who don't have either brown hair, blue eyes, or both then how many kids have both brown hair and blue eyes?

DO YOUR FIGURING HERE

14. If Set A contains (M, N, O, P, Q, R) and Set B contains (K, L, M, N, O, P, Q), how many values are in the intersection of Set A and Set B?

15. In a survey, 95 moviegoers offered their opinions about movie A and movie B. Of those moviegoers, 5 liked neither movie A nor movie B, while 12 liked both movie A and movie B. If 48 moviegoers liked movie A only, how many liked movie B only?

    A. 12

    B. 20

    C. 30

    D. 35

    E. 42

16. All cookies in an assortment contain either chocolate chips, hazelnuts, or both. If 30 cookies contain chocolate chips, 26 cookies contain hazelnuts, and 15 cookies of the cookies contain chocolate chips but no hazelnuts then how many cookies contain hazelnuts but not chocolate chips?

17. There are 50 boxes in a warehouse. Each box contains a notebook, a pencil or both a notebook and a pencil. If 40 of the boxes contain notebooks and 24 of them contain pencils then how many of the boxes contain both a notebook and a pencil?

Venn Diagrams, Unions, and Intersections
Algebra and Arithmetic Problem Set 23

18. At Byram Hills High School, the chess club has 24 members and the computer club has 32 members. If a total of 44 students belong to only one of the two clubs then how many students belong to both clubs?

F. 2

G. 4

H. 6

J. 8

K. 18

19. In a survey, 120 people were asked about whether they played video games A and B. Of the people surveyed, 77 said they play game A, 81 play game B and 12 play neither game. How many people surveyed play both games?

A. 27

B. 31

C. 39

D. 43

E. 50

20. At a certain school, students are required to take chemistry, physics, or both. A total of 13 students take physics, 18 students take chemistry, and a total of 21 students take exactly one science class. How many students take both chemistry and physics?

F. 5

G. 8

H. 13

J. 18

K. 26

DO YOUR FIGURING HERE

21. Each athlete in a group of 50 athletes plays tennis, golf, or both. The total number of athletes playing golf is seven more than the total number of athletes playing tennis. If the number of students that play both sports is the same as the number playing exactly one sport, how many athletes in the group play only golf?

   A. 6

   B. 9

   C. 16

   D. 21

   E. 25

DO YOUR FIGURING HERE

Venn Diagrams, Unions, and Intersections
Algebra and Arithmetic Problem Set 23

Answer Key

#	Answer	Frequency	Difficulty
1	C	popular	1
2	H	popular	2
3	E	popular	2
4	F	popular	2
5	E	popular	3
6	$\{-3, 3, 4, 5, 6, 7, 8, 9, 12\}$	popular	1
7	$\{-3, 6, 9\}$	popular	1
8	8	popular	2
9	$\{-2, 4\}$	popular	1
10	F	popular	1
11	B	popular	3
12	110	popular	3
13	4	popular	3
14	5	popular	2
15	C	popular	2
16	11	popular	2
17	14	popular	2
18	H	popular	4
19	E	popular	3
20	F	popular	4
21	C	popular	4

Direct and Inverse Variation
Algebra and Arithmetic Problem Set 24

1. M varies directly with P. When M is 15, P is 5. What is M when P is 20?

DO YOUR FIGURING HERE

2. y varies inversely with x. When y is 4, x is 25. What is y when x is 10?

3. R varies directly with the square of T. When R is 242, T is 11. What is R when T is 15?

4. D varies directly with the cube root of Q. When D is 2, Q is 27. What is the value of D when Q is 1,728?

Direct and Inverse Variation
Algebra and Arithmetic Problem Set 24

. The values of a and b in the table below are related so that $(b+3)$ is directly proportional to $(a-3)$. What is the value of $n-m$?

a	b
0	6
m	3
n	0

DO YOUR FIGURING HERE

. The price of coffee beans is directly proportional to the weight of the coffee beans in pounds. A 3-pound bag costs \$28.00. What is the cost, in dollars, of p pounds of this coffee.

F. $\dfrac{3}{28}p$

G. $\dfrac{3}{8}p$

H. $\dfrac{8}{3}p$

J. $3p$

K. $\dfrac{28}{3}p$

. If $n \neq 0$ and n is inversely proportional to r, which of the following is directly proportional to $\dfrac{1}{n^3}$?

A. r^3

B. r

C. $r^{\frac{1}{3}}$

D. $\dfrac{1}{r}$

E. $\dfrac{1}{r^3}$

8. If $y = \dfrac{h}{x}$, where h is a constant, and if $y = 2$ when $x = 6$, what does y when $x = 3$?

   F. 1

   G. 3

   H. 4

   J. 6

   K. 12

DO YOUR FIGURING HERE

9. In math classs, Mr. Scone says: "a varies directly as the product of b^2 and c, and inversely as d^3." Which of the following equations, with k as the constant of proportionality, is a correct translation of Mr. Scone's statement?

   A. $a = \dfrac{kb^2c}{d^3}$

   B. $a = \dfrac{kd^3}{b^2c}$

   C. $a = \dfrac{b^2cd^3}{k}$

   D. $a = \dfrac{d^3}{kb^2c}$

   E. $a = kb^2cd^3$

Direct and Inverse Variation
Algebra and Arithmetic Problem Set 24

Answer Key

#	Answer	Frequency	Difficulty
1	60	average	1
2	10	average	3
3	450	average	3
4	8	average	3
5	1	average	3
6	K	average	1
7	A	average	3
8	H	average	1
9	A	popular	2

Comparing Arithmetic and Geometric Sequences
Quick Drill

. Determine if the sequence is arithmetic, geometric, or neither.
2, 6, 10, 14, 18, ...

5. Determine if the sequence is arithmetic, geometric, or neither.
9, 25, 49, 81, 121, ...

. Determine if the sequence is arithmetic, geometric, or neither.
36, 42, 48, 54, 60, ...

6. Determine if the sequence is arithmetic, geometric, or neither.
-64, -59, -54, -49, -44, ...

. Determine if the sequence is arithmetic, geometric, or neither.
$7, \dfrac{5}{12}, \dfrac{7}{12}, 8, \dfrac{8}{13}, ...$

7. Determine if the sequence is arithmetic, geometric, or neither.
2, 6, 18, 54, 162, ...

8. Determine if the sequence is arithmetic, geometric, or neither.
$1, \dfrac{1}{5}, \dfrac{1}{25}, \dfrac{1}{125}, \dfrac{1}{625}, ...$

. Determine if the sequence is arithmetic, geometric, or neither.
-9, 18, -36, 72, -144, ...

Comparing Arithmetic and Geometric Sequences
Quick Drill

9. Determine if the sequence is arithmetic, geometric, or neither.

$-243, 81, -27, 9, -3, ...$

14. Determine if the sequence is arithmetic, geometric, or neither.

$$a_n = -\frac{3}{5} + \frac{4}{3}n$$

10. Determine if the sequence is arithmetic, geometric, or neither.

$1, 2, 5, 8, 13, ...$

15. Determine if the sequence is arithmetic, geometric, or neither.

$$a_n = 17 + 12n$$

11. Determine if the sequence is arithmetic, geometric, or neither.

$a_n = -174 + 50n$

16. Determine if the sequence is arithmetic, geometric, or neither.

$$a_n = \left(\frac{3}{4}n\right)^3$$

12. Determine if the sequence is arithmetic, geometric, or neither.

$a_n = 12 - 4n$

17. Determine if the sequence is arithmetic, geometric, or neither.

$a_n = -57 + 3n$

13. Determine if the sequence is arithmetic, geometric, or neither.

$a_n = (\text{-}2)\left((\text{-}7)^{n+1}\right)$

Comparing Arithmetic and Geometric Sequences
Quick Drill

18. Determine if the sequence is arithmetic, geometric, or neither.

$a_n = \dfrac{3n}{3^n}$

22. Determine if the sequence is arithmetic, geometric, or neither.

$a_n = (4n)(a_{n-1})$
$a_1 = 13$

19. Determine if the sequence is arithmetic, geometric, or neither.

$a_n = -(-6)^{n+1}$

23. Determine if the sequence is arithmetic, geometric, or neither.

$a_n = (a_{n-1})(-22)$
$a_1 = 5$

20. Determine if the sequence is arithmetic, geometric, or neither.

$a_n = (15)\left((-8)^{n-1}\right)$

24. Determine if the sequence is arithmetic, geometric, or neither.

$a_n = a_{n-1} + 31$
$a_1 = -44$

21. Determine if the sequence is arithmetic, geometric, or neither.

$a_n = a_{n-1} + 14$
$a_1 = -3$

25. Determine if the sequence is arithmetic, geometric, or neither.

$a_n = \dfrac{16 + 2a_{n-1}}{4}$
$a_1 = -9$

Comparing Arithmetic and Geometric Sequences
Quick Drill

26. Determine if the sequence is arithmetic, geometric, or neither.

$$a_n = a_{n-1} - 50$$
$$a_1 = 11$$

27. Determine if the sequence is arithmetic, geometric, or neither.

$$a_n = a_{n-1} + 63$$
$$a_1 = -49$$

28. Determine if the sequence is arithmetic, geometric, or neither.

$$a_n = (3n)(a_{n-1})$$
$$a_1 = 4$$

Comparing Arithmetic and Geometric Sequences
Quick Drill

Answer Key

#	Answer
1	Arithmetic
2	Arithmetic
3	Neither
4	Geometric
5	Neither
6	Arithmetic
7	Geometric
8	Geometric
9	Geometric
10	Neither
11	Arithmetic
12	Arithmetic
13	Geometric
14	Arithmetic
15	Arithmetic
16	Neither
17	Arithmetic
18	Neither
19	Geometric
20	Geometric
21	Arithmetic
22	Neither
23	Geometric
24	Arithmetic
25	Neither
26	Arithmetic
27	Arithmetic
28	Neither

Consecutive Numbers
Algebra and Arithmetic Problem Set 25

1. If a represents an even integer, which of the following represents the next even integer greater than a?

   A. $a - 2$

   B. $a - 1$

   C. $a + 1$

   D. $a + 2$

   E. $2a + 1$

DO YOUR FIGURING HERE

2. The sum of four consecutive odd integers is 624. If x represents the least of the four integers, which of the following equations represents the statement above?

   F. $4x + 3 = 624$

   G. $4x + 6 = 624$

   H. $4x + 12 = 624$

   J. $4x + 18 = 624$

   K. $4x + 24 = 624$

3. What is the greatest of 7 consecutive integers if the sum of these integers equals 231?

4. The median of a set of 7 consecutive integers is 78. What is the smallest of these 7 integers?

Consecutive Numbers
Algebra and Arithmetic Problem Set 25

Four consecutive integers are such that one half the largest is two less than the smallest. What is the smallest of these four integers?

A. 5

B. 6

C. 7

D. 8

E. 9

If n and m are positive consecutive even integers, where $n < m$, which of the following is equal to $n^2 - m^2$?

F. $4n + 2$

G. $4n + 4$

H. $-4n - 4$

J. -4

K. $-4n$

A positive integer is said to be "bi-factorable" if it is the product of two consecutive even integers. How many positive integers less than 100 are bi-factorable?

If n is the sum of 3 consecutive integers and m is one of the 3 integers, which of the following could be true?

F. $n = 3m - 2$

G. $n = 3m - 1$

H. $n = 3m$

J. $n = 3m + 6$

K. $n = 3m + 12$

Consecutive Numbers
Algebra and Arithmetic Problem Set 25

Answer Key

#	Answer	Frequency	Difficulty
1	D	popular	1
2	H	popular	3
3	36	popular	3
4	75	popular	2
5	C	popular	3
6	H	popular	3
7	4	popular	3
8	H	popular	3

Arithmetic and Geometric Sequences
Algebra and Arithmetic Problem Set 26

. The first term of an arithmetic sequence is 20 and the second term is 14. What is the third term?

   A. 2

   B. 8

   C. 9.8

   D. 34

   E. 280

DO YOUR FIGURING HERE

· What is the 13th term of an arithmetic sequence whose first term is 91 and whose common difference is -21?

   F. -273

   G. -182

   H. -161

   J. -112

   K. 161

. The first term of an arithmetic sequence is 1 and the third term is 9; what is the sum of the first 9 terms of the sequence?

   A. 153

   B. 171

   C. 180

   D. 189

   E. 234

. What is the sum of the first five terms of the arithmetic series with the first term of 2 and the general term formula for the nth term of $5n-3$?

   F. $\dfrac{1}{2}$

   G. 8

   H. 52

   J. 60

   K. 77

5. What is the sum of the first 128 terms of the arithmetic sequence that has an initial term of 0 and a common difference of 2?

   A. 256

   B. 512

   C. 16,256

   D. 16,384

   E. 16,512

6. Which of the following is NOT true about the arithmetic sequence 1, -2, -5, -8...?

   F. Its fifth term is -11

   G. The sum of its first four terms is -14

   H. The common difference between terms is -3

   J. Its tenth term is -26

   K. The common ratio between terms is -3

7. A certain store charges $2 for a bagel, but charges five cents less for a second bagel, and five cents less than that for a third and so on (so one bagel is $2, a second is $1.95, a third is $1.90). What is the price of 6 bagels?

   A. $10

   B. $10.25

   C. $11

   D. $11.25

   E. $12

8. On day 1, George puts 1 penny in his piggy bank, on day 2, he puts 2 pennies in, on day 3, he puts three pennies in and so on. How many dollars will George have in his piggy bank on Day 365?

   F. $3.65

   G. $36.50

   H. $664.30

   J. $667.95

   K. $66,975.00

DO YOUR FIGURING HERE

Arithmetic and Geometric Sequences
Algebra and Arithmetic Problem Set 26

Andrew scores one run on his first day of practice. On each subsequent day of practice, he scores an additional 3 runs. How many runs did he score in total in his first 5 days?

A. 4

B. 10

C. 13

D. 22

E. 35

DO YOUR FIGURING HERE

. The interior angles of a triangle form an arithmetic sequence with a common difference of 10. What is the degree measure of the largest angle?

F. 50

G. 60

H. 70

J. 80

K. 90

. The second term of an arithmetic sequence is 8 and the ninth term is 29. What is the fifth term?

A. 3

B. 17

C. 18.5

D. 20

E. 23

. The first term of an arithmetic sequence is 3 and the common difference is 2. What is the sum of the first 20 terms?

F. 380

G. 400

H. 440

J. 460

K. 620

13. What is the next term in the geometric sequence:
24.2, 14.52, 8.712. . .? (Round to the nearest hundredth)

A. -0.97

B. 0.00

C. 1.00

D. 2.90

E. 5.23

DO YOUR FIGURING HERE

14. The terms of a geometric sequence are $\frac{1}{3}, 1, 3, x,$ 27. What is the value of x?

F. 3

G. 6

H. 9

J. 12

K. 15

15. What is the value of x in the geometric sequence $x, 9, 6, 4, \frac{8}{3} \ldots$?

A. $\frac{3}{2}$

B. 12

C. $\frac{27}{2}$

D. 18

E. 27

16. The second term of a geometric sequence is 2 and the fifth term is 0.016. What is the first term of the sequence?

F. 0.2

G. 0.4

H. 10

J. 20

K. 50

Arithmetic and Geometric Sequences
Algebra and Arithmetic Problem Set 26

7. The sum of an infinite geometric sequence with the first term a and ratio r is $\frac{a}{(1-r)}$. If the sum of an infinite geometric sequence is 500 and its common ratio is .25, what is its second term?

   A. 24.4375

   B. 93.75

   C. 125

   D. 281.25

   E. 375

8. The sum of an infinite geometric sequence with the first term a and ratio r is $\frac{a}{(1-r)}$. What is the sum of an infinite geometric sequence that starts out with $10, 8, 6.4, 5.12$?

   F. 0.2

   G. 0.8

   H. 12.5

   J. 29.52

   K. 50

9. The sum of an infinite geometric sequence with the first term a and ratio r is $\frac{a}{(1-r)}$. The sum of an infinite geometric sequence is 40 and its first term is 20. What is the common ratio between its terms?

   A. $\frac{1}{6}$

   B. $\frac{1}{5}$

   C. $\frac{1}{4}$

   D. $\frac{1}{3}$

   E. $\frac{1}{2}$

DO YOUR FIGURING HERE

Arithmetic and Geometric Sequences
Algebra and Arithmetic Problem Set 26

20. The sum of an infinite geometric sequence with the first term a and ratio r is $\dfrac{a}{(1-r)}$. The sum of an infinite geometric sequence is 100 and its common ratio $\dfrac{1}{4}$; what is the first term?

   F. 10

   G. 25

   H. 50

   J. 75

   K. 80

21. The sum of an infinite geometric sequence with the first term a and ratio r is $\dfrac{a}{(1-r)}$. The second term of a geometic sequence is 5 and the common ratio is $\dfrac{1}{2}$; what is the sum of the infinite series?

   A. 5

   B. 10

   C. 20

   D. 50

   E. 100

22. The sum of an infinite geometric sequence with the first term a and ratio r is $\dfrac{a}{(1-r)}$. The fifth term of a geometic sequence with a common ratio of .1 is .04. What is the sum of the infinite geometric sequence?

   F. 400

   G. 444.444

   H. 4,000

   J. 40,000

   K. Cannot be determined

DO YOUR FIGURING HERE

Arithmetic and Geometric Sequences
Algebra and Arithmetic Problem Set 26

DO YOUR FIGURING HERE

3. On the first day of July, Sarah harvests 3 zucchini. Each day, she harvests 2 more zucchini than she did on the previous day. How many zucchini did she harvest on July 15?

- **A.** 15 zucchini
- **B.** 21 zucchini
- **C.** 29 zucchini
- **D.** 31 zucchini
- **E.** 33 zucchini

4. Sam is training for a marathon. Each Saturday, she runs 2 more miles than she ran on the previous Saturday. If she runs 4 miles on the first week, on what week will she go on a 26-mile run?

- **F.** Week 10
- **G.** Week 11
- **H.** Week 12
- **J.** Week 14
- **K.** Week 15

5. Jane reads 3 pages of a mystery novel on the first day. Each day, she reads 1.5 times as many pages as on the previous day. How many total pages of her book has she read by the end of the 5th day?

- **A.** 22.78 pages
- **B.** 33.25 pages
- **C.** 39.56 pages
- **D.** 87.48 pages
- **E.** 91.13 pages

6. In a snowstorm, 1 inch falls in the first hour. Every hour after that, twice as much snow falls than fell in the previous hour. During which hour does the snow level reach 1 foot in total?

- **F.** Hour 2
- **G.** Hour 3
- **H.** Hour 4
- **J.** Hour 5
- **K.** Hour 6

Arithmetic and Geometric Sequences
Algebra and Arithmetic Problem Set 26

Answer Key

#	Answer	Frequency	Difficulty
1	B	popular	1
2	H	popular	2
3	A	popular	3
4	J	popular	3
5	C	popular	2
6	K	popular	2
7	D	popular	1
8	J	popular	3
9	E	popular	2
10	H	popular	2
11	B	popular	3
12	H	popular	2
13	E	popular	2
14	H	popular	2
15	C	popular	3
16	H	popular	4
17	B	popular	3
18	K	popular	3
19	E	popular	3
20	J	popular	3
21	C	popular	4
22	G	popular	5
23	D	popular	5
24	H	popular	5
25	C	popular	5
26	H	popular	5

Patterns and Remainders
Algebra and Arithmetic Problem Set 27

. What is the remainder when 12 is divided by 7?

. What is the remainder when 113 is divided by 8?

. The first term of a series of numbers is 6. If each term after the first term equals 6 more than half the previous term, what is the sixth term in this series?

. Whenever Paul goes to the sports store he buys as many basketball cards as he can afford. If Paul walks into the sports store with $100 and each pack of cards he buys are $6, how much money will Paul have left when he leaves the store?

. If today is Sunday, what day will it be 32 days from now?

   A. Tuesday

   B. Wednesday

   C. Thursday

   D. Saturday

   E. Monday

. Linda is transferring 1 gallon of orange juice into 7 oz. bottles so she can take a small bottle to school each day. If there are approximately 128 oz. in 1 gallon, how many ounces of orange juice will be left over if Linda only wants full 7 oz. bottles?

   F. 1

   G. 2

   H. 3

   J. 4

   K. 18

Patterns and Remainders
Algebra and Arithmetic Problem Set 27

7. A sequence is formed by repeating the numbers 4, -7, 3, and 2 over and over in the same order. If this pattern is repeated until there are 40 terms in the sequence, what is the sum of these terms?

DO YOUR FIGURING HERE

8. At the grand opening of a new supermarket, every sixth customer gets a free chicken wing and every fifteenth customer receives a free soy milk for the first 500 customers. How many customers will receive a door prize?

9. The first three numbers of a pattern are 4, 11, and 18. Each term after the first three is found by taking the average of all the preceding terms. What is the 33rd term in the sequence?

    A. 11

    B. 76

    C. 198

    D. 228

    E. 235

10. The first term of a sequence is 5. Each term after the first is found by subtracting 2 from the previous term and then multiplying it by -1. What is the 45th term of the sequence?

    F. -81

    G. -5

    H. -3

    J. 3

    K. 5

2019-1

Patterns and Remainders
Algebra and Arithmetic Problem Set 27

. A beaded necklace repeats the following pattern: 3 red beads, 2 green beads, 1 black bead, 2 green beads. How many red beads would be used to make a necklace with 196 beads?

A. 70

B. 72

C. 74

D. 75

E. 78

. At JFK airport, every third piece of luggage is placed through the metal detector and every 15th piece of luggage is inspected. If 8,400 pieces of luggage pass through JFK in one day, how many are both inspected and put through the metal detector?

F. 186

G. 280

H. 560

J. 1,800

K. 2,800

Answer Key

#	Answer	Frequency	Difficulty
1	5	average	1
2	1	average	1
3	11.8125	average	3
4	$4	average	2
5	C	average	2
6	G	average	2
7	20	average	2
8	100	average	5
9	A	average	1
10	K	average	2
11	D	average	3
12	H	average	2

Logical Reasoning
Algebra and Arithmetic Problem Set 28

Mikala, Simba and Stefan all accidentally took one another's backpack home from school. In how many ways can they be paired with a backpack so that no one has his own?

A. 1

B. 2

C. 3

D. 6

E. 8

DO YOUR FIGURING HERE

How many positive three digit integers have the tens place equal 2 and the ones digit equal 1?

F. 1

G. 2

H. 3

J. 9

K. 10

The five children in the Turkel family are Bo, Clarence, Alejandro, Ronald and Hans. Bo and Ronald are both neither youngest nor oldest. Clarence is the middle child. Bo and Alejandro both have more younger siblings than Ronald. Who is the oldest?

A. Alejandro

B. Bo

C. Clarence

D. Hans

E. Ronald

Adrianna is older than Yanna but younger than Colby. A, Y and C represent their ages, respectively. Which of the following is true?

F. $A < Y < C$

G. $A < C < Y$

H. $C < Y < A$

J. $Y < C < A$

K. $Y < A < C$

Logical Reasoning
Algebra and Arithmetic Problem Set 28

5. There are 5 red, 5 brown, 5 green and 5 purple sodas. Each soda is in its own box and all the boxes are identical. What is the least number of boxes that need to be opened to be sure that 3 or more opened boxes contain the same color soda?

    A. 8

    B. 9

    C. 10

    D. 13

    E. 20

6. Each digit is 3, 5, 7, or 9. Exactly 2 digits are the same. No 2 adjacent digits are the same. How many positive 3-digit integers satisfy all three conditions above?

    F. 3!

    G. 4!

    H. 10

    J. 12

    K. 99

7. Poker chips can either be worth 2, 10, or 20 dollars. How many combinations of chips equal 68 dollars?

    A. 4

    B. 8

    C. 16

    D. 32

    E. 58

8. From a jar containing 30 cookies, of which 15 are chocolate chip and 15 are oatmeal, Emilio has taken 4 oatmeal and 2 chocolate chip. He takes an additional 12 cookies. What is the least number of these additional 12 cookies that Emilio takes that must be chocolate chip in order for him to have more chocolate chip than oatmeal?

Logical Reasoning
Algebra and Arithmetic Problem Set 28

Answer Key

#	Answer	Frequency	Difficulty
1	B	rare	2
2	J	rare	2
3	A	average	3
4	K	rare	2
5	B	rare	3
6	J	average	3
7	C	rare	4
8	8	rare	2

Formulas
Quick Drill

1. What is the area of a rectangle?

2. What is the area of a square?

3. What is the volume of a rectangular prism?

4. What is the volume of a cube?

5. What is the surface area of a rectangular prism?

6. What is the surface area of a cube?

7. What is the area of a circle?

8. What is the circumference of a circle?

9. What is the volume of a cylinder?

10. What is the surface area of a cylinder?

11. What is the distance formula?

12. What is the 3-dimensional distance formula?

13. What is the midpoint formula?

14. What is the equation of a line in slope-intercept form?

15. What does m stand for and what is the formula for it?

16. What does b stand for?

17. What is the area of a triangle?

18. What is the formula for finding the hypotenuse of of a right triangle? What is it called?

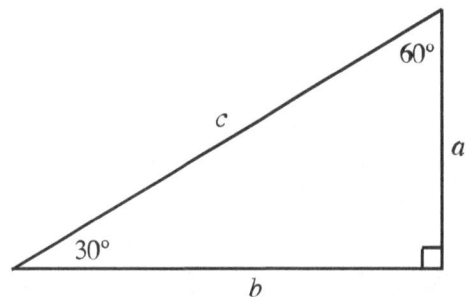

19. Label all sides in the 30-60-90 triangle.

Formulas
Quick Drill

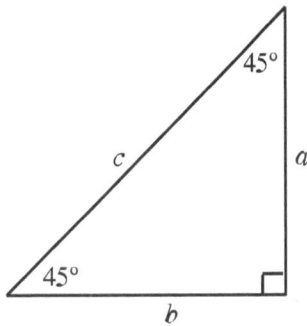

20. Label all sides in the 45-45-90 triangle.

21. What is the area of a parallelogram?

22. What is the area of a trapezoid?

23. What is the sum of the measure of the interior angles of a polygon?

24. What is the measure of one interior angle of a regular polygon?

25. What is the quadratic formula?

26. What is the formula for arithmetic mean?

27. What is a median?

28. What is a mode?

29. What is the formula for the common difference of an arithmetic sequence?

30. What is the formula for the nth term of an arithmetic sequence?

31. What is the formula for the sum of n terms of an arithmetic sequence?

32. What is the formula for the common ratio of a geometric sequence?

33. What is the formula for translating radians into degrees?

34. What is the equation for a circle?

35. What is the equation for an ellipse?

36. What does SOHCAHTOA stand for?

Formulas
Quick Drill

Answer Key

#	Answer
1	$A = lw$
2	$A = s^2$
3	lwh
4	$V = s^3$
5	$2\left(lw + hl + hw\right)$
6	$SA = 6s^2$
7	$A = \pi r^2$
8	$C = \pi d$
9	$V = \pi r^2 h$
10	$SA = 2\pi r^2 + \pi dh$
11	distance = $\sqrt{\left(x_2 - x_1\right)^2 + \left(y_2 - y_1\right)^2}$
12	distance = $\sqrt{\left(x_2 - x_1\right)^2 + \left(y_2 - y_1\right)^2 + \left(z_2 - z_1\right)^2}$
13	midpoint = $\left(\dfrac{x_1 + x_2}{2} + \dfrac{y_1 + y_2}{2}\right)$
14	$y = mx + b$
15	slope = $\dfrac{y_2 - y_1}{x_2 - x_1}$
16	y-intercept
17	$A = \dfrac{bh}{2}$
18	$a^2 + b^2 = c^2$, Pythagorean Theorem
19	$a = x$, $b = x\sqrt{3}$, $c = 2x$
20	$a = x$, $b = x$, $c = x\sqrt{2}$
21	$A = bh$
22	$A = \dfrac{h\left(b_1 + b_2\right)}{2}$
23	$180\left(n - 2\right)$
24	$\dfrac{180\left(n - 2\right)}{n}$
25	$x = \dfrac{-b \pm \sqrt{b^2 - 4ac}}{2a}$
26	$\dfrac{1}{n}\displaystyle\sum_{i=1}^{n} x_i$
27	the middle number in a set in numerical order

#	Answer
28	the most common number in a set
29	$d = a_n - a_{n-1}$
30	$a_n = a_1 + \left(n - 1\right)d$
31	$\displaystyle\sum_{i=1}^{n} a_i = \dfrac{n}{2}\left(a_1 + \boldsymbol{a_n}\right)$
32	$r = \dfrac{a_n}{a_{n-1}}$
33	degree = radian$\left(\dfrac{180}{\pi}\right)$
34	$\left(x - h\right)^2 + \left(y - k\right)^2 = r^2$, where $\left(h, k\right)$ is the center
35	$\dfrac{\left(x - h\right)^2}{a^2} + \dfrac{\left(y - k\right)^2}{b}$ where $\left(h, k\right)$ is the center and $2a$ and $2b$ are the lengths of the axes
36	sine=$\dfrac{\text{opposite}}{\text{hypotenuse}}$, cosine=$\dfrac{\text{adjacent}}{\text{hypotenuse}}$, and tangent=$\dfrac{\text{opposite}}{\text{adjacent}}$